Lecture Notes in Physics

Volume 935

The Lecture Notes in Physics

The series Lecture Notes in Physics (LNP), founded in 1969, reports new developments in physics research and teaching-quickly and informally, but with a high quality and the explicit aim to summarize and communicate current knowledge in an accessible way. Books published in this series are conceived as bridging material between advanced graduate textbooks and the forefront of research and to serve three purposes:

- to be a compact and modern up-to-date source of reference on a well-defined topic
- to serve as an accessible introduction to the field to postgraduate students and nonspecialist researchers from related areas
- to be a source of advanced teaching material for specialized seminars, courses and schools

Both monographs and multi-author volumes will be considered for publication. Edited volumes should, however, consist of a very limited number of contributions only. Proceedings will not be considered for LNP.

More information about this series at http://www.springer.com/series/5304

Yoshiaki Sofue

Galactic Radio Astronomy

Springer

Yoshiaki Sofue
Institute of Astronomy
The University of Tokyo
Mitaka, Tokyo, Japan

ISSN 0075-8450 ISSN 1616-6361 (electronic)
Lecture Notes in Physics
ISBN 978-981-10-3444-2 ISBN 978-981-10-3445-9 (eBook)
DOI 10.1007/978-981-10-3445-9

Library of Congress Control Number: 2017940277

Printed on acid-free paper

This Springer imprint is published by Springer Nature
The registered company is Springer Nature Singapore Pte Ltd.
The registered company address is: 152 Beach Road, #21-01/04 Gateway East, Singapore 189721, Singapore

Abstract

Chapter 1

The Milky Way Galaxy, in which we live, is a giant 20 kpc diameter disk. The center of the galaxy lies about 8 kpc away from the Sun in the constellation Sagittarius. The interstellar matter (ISM) contains a significant amount of dust that hides most of the optical objects in the galactic disk from us. The optical absorption rate in the solar vicinity is about 1 mag per 1 kpc, and the galactic extinction amounts to $\sim$30 mag for the light from the galactic center (GC), or the light is weakened by $\sim 100^{-(1/5)20\sim 30} = 10^{-8\sim -12}$ times the original intensity. Namely, the GC cannot be observed in visible light, even if we used the world largest telescopes. The GC and the regions beyond are a world hidden from our eyes.

These deepest parts of the galaxy can be easily visible if we observe them through the radio window. Radio waves are transparent against the interstellar dust because they have much longer wavelengths than the dust, and avoid scattering by dust particles. This situation mimics the reception of radio and TV waves even on a foggy and rainy day. If it is seen in radio, the radio galaxy appears quite different from its optical appearance.

The change in appearance is not only because of the difference in opacity, but also because of the different locations and matter that emit radiation as well as because of the emission mechanism. Knowledge about the emission mechanism can in turn be used to investigate the origin and physics of the emitting regions.

In this chapter the fundamental physics of various emission mechanisms in interstellar space is described, starting with basic formulae for electromagnetic waves. Readers who feel this chapter is too basic may skip here and proceed to the chapters more concerned with astrophysics and astronomy, and may come back when more fundamental knowledge is necessary to understand the underlying physics of the phenomena. In this book, we use the cgs (cm, gram, second) unit system.

Chapter 2

Interstellar space is filled with interstellar matter (ISM) composed of various kinds of gases from low temperature (molecules, HI gas, and dust) to high temperature (ionized gas, HII regions), high-energy particles (cosmic rays), and magnetic fields. They are embedded in the radiation field of stars. The ISM makes up approximately

10% of the mass of the galactic disk, and the majority is HI and H_2. The remaining 90% is shared by stars and dark matter, which determine the dynamical properties of the galaxy. The IMS is deeply coupled to activities such as the star formation near the galactic plane, and is categorized as younger-generation matter in the galaxy, called Population I.

Chapter 3

The main body of the luminous disk of the galaxy is composed of stars that were born when the galaxy was formed about 13 billion years ago. These first-generation, old stars are today observed in globular clusters, the thick galactic disk, and the central bulge. The Sun was born about 4.6 billion years ago in one of the molecular clouds near the galactic plane.

Some portion of the interstellar gas has survived from being exhausted in the formation of the stars, but has been polluted by contamination of ejected gas from the stars with higher metallicity through stellar winds and supernova explosions. Star formation is thus taking place in the galaxy even today from the remaining gas mostly distributed in the thin, dense gaseous disk near the galactic plane. Newly born, young, and bright stars illuminate and excite surrounding clouds by radiation at various wavelengths, from radio to X-rays. Star- forming regions are thus bright among various objects, and are relatively easily observed and investigated in detail.

Chapter 4

The Milky Way is one of the most typical spiral galaxies, resembling the Andromeda galaxy (M31), and is composed of a nucleus nesting a supermassive black hole, a spheroidal central bulge composed of old stars, flat stellar disk filled with interstellar gas superposed by spiral arms, and a huge dark halo extending half-way to the neighboring galaxy.

The galaxy is the nearest, hence most extensively observed galaxy in detail, however, the special position of the observers located inside the object always poses difficulty in knowing the three-dimensional structure. This difficulty, in turn, forced astronomers to develop sophisticated techniques of observation and analysis. The difficulty is overcome by observing a galaxy from outside, which is in fact extensively done for external galaxies.

In this chapter we describe the method to derive the dynamical structure of galaxies based on the Newtonian theory of gravitation. We learn various methods to measure the mass distributions, and the methods are applied to observed rotation curves of galaxies.

Chapter 5

The galactic center has been a special object in astronomy since its discovery by C. Jansky as one of the brightest radio sources. It is a particularly useful object for understanding the nuclear activities of galaxies and their central black holes because of the small distance from us.

Although nuclei of galaxies are not identical to the galactic center (GC), their fundamental physics and exhibited phenomena are in general similar, because the galactic structure and the composing material are similar, or sometimes almost the same. Hence, the galactic center and nuclear regions of galaxies are in general essentially the same. In this chapter various radio observational phenomena in the galactic center are reviewed and compared with those in active nuclei in galaxies. They are discussed in terms of energetic activities in the nuclear regions such as explosions and starbursts.

Chapter 6

Radio observations of the synchrotron radiation from the galactic disk and its polarization give us information about the magnetic field structure of the galaxy. Analyses of polarization vectors and of Faraday rotation measures along the line of sight give us information about the three-dimensional orientation of magnetic fields.

Magnetic fields are in pressure equilibrium with the interstellar medium on the order of $\sim 10^{-12}$ ergs cm^{-3}. These fields are coupled with interstellar gas, often termed "frozen in" to each other, and influence the hydrodynamic structure of the galactic disk.

In the central regions, the magnetic fields accumulate to create strong vertical fields, which are wound by galactic rotation, and produce cosmic jets from the nuclear disks, often exhibiting highlighted cosmic jets from active galactic nuclei nesting supermassive black holes.

The strongest, and most compact, magnetic field thus far observed reaches $\sim 10^{12}$ G on the surface of neutron stars observed as pulsars. Galactic magnetic fields are on the order of several 10^{-6} G or ~ 1 μG. Intergalactic fields are sometimes observed to exhibit about the same order as galactic, ~ 1 μG.

The largest-scale, but weaker, magnetic field may penetrate the universe. Thus far known celestial objects are all magnetized, from the Earth, the Sun, stars, interstellar matter, galaxies, and clusters of galaxies. Only one exception is the universe for which the cosmological isotropy principle is adopted. It might be rather challenging to ask if the universe has the north and south directed by a cosmological magnetic field.

Preface

The Solar system—the Sun, the Earth, and planets—is a tiny family in the Galaxy as one of the numerous similar families of stars. The Galaxy consists of stars, gas, and dark matter totaling some 10^{12} solar masses, and rotates around the Galactic Center in 200 million years. The Sun was born 4.6 billion years ago in a local gas cloud 25,000 light years away from the Galactic Center, when the Galaxy was 9 billion years old. Stars are being born even today in interstellar clouds, making the galactic disk and spiral arms active as star-forming sites. The Galactic Center is the most active region with bursting star formation, exhibiting explosive phenomena, and nests a super-massive black hole of four million solar masses. The Galaxy is one of the numerous galaxies in the universe, a typical spiral similar to the beautiful neighbor, the Andromeda galaxy.

These dazzling galactic objects, sometimes revealing apparently mysterious phenomena, are understood and described by simple physics. All the objects are observed by the electromagnetic waves radiated by moving electrons. The radiation mechanisms are the established fields of physics, clearly understood and described by electromagnetic theory. This means that any celestial objects, emitting radiation, are not mysterious zones at all, but are describable by fundamental physics. The emissions from the objects transport fundamental information of the sources. For example, the formation of stars is understood by the gravitational force acting on an interstellar cloud, whose density, size, mass, temperature, and so on are measured by observations. The rotation of the Galaxy, as observed using the Doppler effect, is useful to measure the galactic mass based on mechanics.

This book aims at describing the galactic and interstellar phenomena in terms of simple physical laws, and at understanding the underlying physics of the driving mechanisms of the universe.

Tokyo, Japan
August 2016

Yoshiaki Sofue

Contents

Acronyms

List of abbreviations, symbols and the like.

AGN	Active Galactic Nucleus
AGB	Asymptotic Giant Branch
ALMA	Atacama Large mm and submm Array
APOD	Astronomy Picture of the Day
ASS	Axisymmetric Spiral
BH	Black Hole
BBR	Black Body Radiation
BHS	Biploar Hyper Shell
BSS	Bisymmetric Spiral
CBR	Cosmological Background Radiation
CMZ	Central Molecular Zone
CMB	Cosmological Microwave Background
CO	Carbon Monoxide
CR	Cosmic Ray
DH	Dark Halo
DM	Dark Matter
EM	Emission Measure
EMR	Expanding Molecular Ring
FB	Fermi Bubble
FF emission	Free-Free emission
FIR	Far Infrared
GC	Galactic Center
GCL	Galactic Center Lobe; Globular Cluster
GMC	Giant Molecular Cloud
HI	Neutral Hydrogen
HII	Ionized Hydrogen
HVC	High-Velocity Cloud
IMF	Initial Mass Function
IR	Infrared
ISM	Interstellar Matter
LG	Local Group

LSB	Low Surface Brightness
LSR	Local Standard of Rest
LVD	Longitude-Velocity Diagram
MC	Molecular Cloud
MHD	Magneto-Hydrodynamics
ML ratio	Mass-to-Luminosity ratio
MW	Milky Way
NGC	New General Catalog
NGP	North Galactic Pole
NPS	North Polar Spur
NRO	Nobeyama Radio Observatory
PN	Planetary Nebula
PVD	Position-Velocity Diagram
RC	Rotation Curve
RM	Rotation Measure
SB	Star Burst
SF	Star Formation
SFR	Star Formation Rate
SGP	South Galactic Pole
SMBH	Super Massive Black Hole
SMD	Surface Mass Density
SN	Supernova
SNR	Supernova Remnant
TFR	Tully-Fisher Relation
UV	Ultra Violet
VERA	VLBI Experiment for Radio Astronometry
VLA	Very Large Array
VLBI	Very Long Baseline Interferometer
WMAP	Wilkinson Microwave Anisotry Probe

The original version of this book was published without printing chapter abstract. The abstracts for each chapter are updated in the front matter in the current version of the book. An erratum to this chapter can be found at https://doi.org/10.1007/978-981-10-3445-9_7.

Chapter 1
Introduction

1.1 Radio Astronomy

1.1.1 A History of Radio Astronomy

The history of astronomy began a few thousand years ago when human beings watched the universe in optical light, for our eyes are sensitive only in optical wavelengths. The new wavelength window was opened only a half-century ago, when C. Jansky detected a radio signal from the galactic center for the first time.

This young radio engineer was listening to noise from the sky, and found a radio source that did not follow the rotation of the Earth relative to the Sun, but followed the celestial rotation. The source is well known today as one of the brightest radio sources in the sky, Sgr. A, the center of our galaxy. Since we opened the radio window toward the universe, we have obtained revolutionary progress in obtaining knowledge of the radio world mainly thanks to the technical development of detectors, telescopes, and data reduction powers. Over the years, particularly in the 1960s and 1970s, some of the most important findings in astronomy were due to radio astronomy, including the discovery of the HI 21-cm line emission from interstellar neutral hydrogen, which allowed us to study the galactic structure (1950–1970s); the discovery of cosmic background black-body radiation (1960s); the discovery of a pulsar and neutron stars (1960s); the discovery rush of interstellar molecules (1960–1970s); and the high-resolution interferometric techniques that revealed the structures of cosmic radio sources.

Today, radio observations give us a huge amount of data both in imaging and spectroscopy, which require highly advanced data processing. After the processing, astronomers obtain high-quality images that are now "visible," and are stimulated by the images to learn more about "invisible" objects.

The original version of this chapter was published without printing chapter abstract. The abstracts for each chapter are updated in the front matter in the current version of the book. An erratum to this chapter can be found at https://doi.org/10.1007/978-981-10-3445-9_7.

Y. Sofue, *Galactic Radio Astronomy*, Lecture Notes in Physics 935,
DOI 10.1007/978-981-10-3445-9_1

1.2 Radio Emission

1.2.1 Plane Waves

In this section we describe the fundamental properties of electromagnetic waves and their propagation [5–7]. The basic equations describing electromagnetic waves in the vacuum are written:

$$\mathrm{rot}\, \boldsymbol{E} = -c\frac{d\boldsymbol{H}}{dt}, \tag{1.1}$$

$$\mathrm{div}\, \boldsymbol{H} = 0, \tag{1.2}$$

$$\mathrm{rot}\, \boldsymbol{H} = \frac{1}{c}\frac{d\boldsymbol{E}}{dt}, \tag{1.3}$$

$$\mathrm{div}\, \boldsymbol{E} = 0, \tag{1.4}$$

where $\boldsymbol{E}$, $\boldsymbol{H}$, and c are the electric and magnetic vectors and the light speed, respectively. Introducing the vector potential $\boldsymbol{A}$ as defined by

$$\boldsymbol{H} = \mathrm{rot}\, \boldsymbol{A}, \tag{1.5}$$

Maxwell's equations can be reduced to the wave equation

$$\Delta \boldsymbol{A} - \frac{\partial^2 \boldsymbol{A}}{c^2 \partial t^2} = 0. \tag{1.6}$$

The electric vector can be related to the magnetic vector as

$$\boldsymbol{E} = \boldsymbol{H} \times \boldsymbol{n}, \tag{1.7}$$

and the vectors can be related to the vector potential as

$$\boldsymbol{H} = \frac{1}{c}\boldsymbol{A} \times \boldsymbol{n}, \tag{1.8}$$

and

$$\boldsymbol{E} = \frac{1}{c}(\boldsymbol{A} \times \boldsymbol{n}) \times \boldsymbol{n}. \tag{1.9}$$

Here, $\boldsymbol{n}$ is the unit vector in the direction of the wave propagation, which is taken in the z direction.

A special solution of the wave equation is given by

$$\boldsymbol{A} = \boldsymbol{a}e^{i(\boldsymbol{k}.\boldsymbol{r}-ft)} \tag{1.10}$$

which represents a wave of wave number $\boldsymbol{k}$ and frequency f. For a wave in the vacuum we have $k/f = c$, and $\boldsymbol{E}$ and $\boldsymbol{H}$ can be written as

$$E_x = E_{x0}\cos(ft - \boldsymbol{k}.\boldsymbol{r} + g), \tag{1.11}$$

and

$$E_y = E_{y0}\sin(ft - \boldsymbol{k}.\boldsymbol{r} + g). \tag{1.12}$$

A wave with $E_{x0} = E_{y0}$ is called a circular polarized wave, a wave with $E_{x0}n_e E_{y0}$ is called elliptically polarized, and a wave with $E_{x0} = 0$ or $E_{y0} = 0$ is called a linearly polarized wave.

The energy carried by an electromagnetic wave whose speed is c is expressed by a Poynting vector

$$\boldsymbol{S} = \frac{c}{4\pi}\boldsymbol{E} \times \boldsymbol{H} = \frac{c}{4\pi}H^2 \times \boldsymbol{n}. \tag{1.13}$$

The Poynting vector can be used to calculate the energy flow per small steradian $d\Omega$ at a distance of r from the radiation source, which is called the flux density:

$$dI = \frac{c}{4\pi}H^2 r^2 d\Omega. \tag{1.14}$$

1.2.2 Acceleration of a Charged Particle and Radiation

Consider a charged particle, for example, an electron. If the particle is at rest, the electric field surrounding this particle does not change. If the particle is moving straight at a constant speed, we can translate the coordinate so that the particle appears to be at rest, and the electric field can also be reduced to a stationary one. Suppose that the particle is moving with acceleration: the simplest case may be a particle oscillating at a frequency f. Then the electric field oscillates at the same frequency and the oscillation is transferred through the vacuum at a speed c.

Because the amplitude of the motion of the particle is maximum when the oscillation is seen perpendicular to the orbit, the electromagnetic waves radiated attain the maximum toward the perpendicular direction. On the other hand, the motion appears be at rest when seen from the line of oscillation, and hence the radiation along the line of oscillation becomes minimum, as the wave is a transverse wave. This kind of radiation of electromagnetic waves is called dipole emission. This situation can be easily understandable if one imagines a straight antenna along which electrons are oscillating: the radiation strength is maximum in the direction perpendicular to the antenna, and is minimum in the direction along it. Suppose an electron whose charge is e with position vector r; the dipole moment is defined by $d = er$. The vector potential is then written as

$$\boldsymbol{A} = \frac{e\dot{\boldsymbol{r}}}{cR} = \frac{\dot{\boldsymbol{d}}}{cR}, \tag{1.15}$$

where R is the distance of a point from the coordinate origin. If the point is far enough from the origin, the waves can be approximated by plane waves. Then E and H of the plane wave are expressed as

$$\boldsymbol{H} = \frac{1}{c^2R}\ddot{\boldsymbol{d}} \times \boldsymbol{n}, \tag{1.16}$$

and

$$\boldsymbol{E} = \frac{1}{c^2R}(\ddot{\boldsymbol{d}} \times \boldsymbol{n}) \times \boldsymbol{n}. \tag{1.17}$$

Let the angle between the wave direction n and the direction of the oscillation (the direction of the vector d) be θ; then the flux density of the radiation from this particle is written as

$$dI = \frac{(\ddot{\boldsymbol{d}} \times \boldsymbol{n})^2}{4\pi c^3} d\Omega = \frac{\ddot{d}^2}{4\pi c^3} \sin^2\theta^2 d\Omega. \tag{1.18}$$

The radiation attains the maximum toward the direction perpendicular to d or to the antenna, and it attains the minimum along the direction of oscillation. By integrating the intensity over the steradian, we can obtain the total intensity I of the radiation from the particle as

$$I = \int dI = \frac{2\ddot{d}^2}{3c^3} = \frac{2e^2\ddot{r}^2}{3c^3}. \tag{1.19}$$

For an oscillation with a frequency f we have $r = r_0\cos(ft - ft_0)$ or $d = d_0\cos(ft - ft_0)$, yielding the total intensity as

$$I = \frac{2d_0^2 f^4}{3c^3}. \tag{1.20}$$

The above expression is for a single electron under a simple oscillation. In nature, however, it is rare that electrons move back and forth in perfect alignment, but are usually more or less in random motion. The acceleration of electrons is due to the orbital motion within an atom, because of collisions with ions in plasma, or the curved motion during the Lorenz interaction with the interstellar magnetic field.

1.2.3 Radiation from an Electron in a Magnetic Field

The simplest acceleration motion of an electron in nature would be a circular motion. For instance, an electron moving through interstellar space suffers from the Lorenz force by the interstellar magnetic field, resulting in a circular (helical) rotation.

The Lorenz force acting on an electron moving at a velocity v in a magnetic field of strength B is given by

$$F = eBv/c. \tag{1.21}$$

This force is in balance with the centrifugal force

$$F = mr\omega^2, \tag{1.22}$$

where m and ω are the mass of the electron and the angular frequency, respectively. Because $v = r\omega$ we have

$$\omega = \frac{eB}{mc}. \tag{1.23}$$

The frequency ω is called the gyro frequency, and the radius of the circular orbit r is the Larmor radius:

$$r = \frac{cmv}{eB} = \frac{v}{\omega}. \tag{1.24}$$

The dipole moment of an electron circularly rotating in a uniform magnetic field can be written as

$$d = er_0 \cos \omega t \tag{1.25}$$

and the intensity of the electromagnetic wave emitted by this electron can be expressed as

$$dI = \frac{e^2 r_0^2 \omega^4}{4\pi c^3} \sin^2\theta d\Omega. \tag{1.26}$$

The total intensity is obtained by integrating the above equation by Ω:

$$I = \frac{2e^2 r_0^2 \omega^4}{3c^3} = \frac{2e^4 B^2 (v/c)^2}{3m^2 c^3}. \tag{1.27}$$

By using the kinetic energy $E = \frac{1}{2}mv^2$ we obtain

$$I = \frac{4e^4B^2}{3m^3c^5}E = c_1E, \tag{1.28}$$

where $c_1 = 4e^4B2/3m^3c^5$ is a constant. Due to the radiation the kinetic energy is lost at a rate of

$$dE/dt = -I = -c_1E. \tag{1.29}$$

Therefore the energy changes as a function of time as

$$E = E_0\exp(-c_1t). \tag{1.30}$$

Here E_0 is the initial kinetic energy of the electron. The decay time of the energy is given by

$$t = 1/c_1 = \frac{3m^3c^5}{4e^4B^2}. \tag{1.31}$$

The above expressions are valid for the case in which the velocity of the electron is small enough compared to the light speed ($v \ll c$).

If the velocity is close to the light speed, or the electron is relativistic, we obtain

$$I = \frac{2e^4B^2\gamma^2}{3m^2c^3} = \frac{2e^4B^2}{3m^4c^7}E^2 = c_2E^2, \tag{1.32}$$

with $\gamma = E/mc^2$. The energy changes as

$$dE/dt = -c_2E^2. \tag{1.33}$$

This gives a solution

$$E = \frac{1}{c_1[t + (1/c_2)E_0]}, \tag{1.34}$$

and the decay time of an electron as

$$t = \frac{1}{c_2E_0} = \frac{3m^4c^7}{2e^4B^2E_0}. \tag{1.35}$$

The radiation expressed by Eqs. (1.26), (1.27), (1.28), (1.29), (1.30), and (1.31) for a nonrelativistic case is called cyclotron radiation. The gradation expressed by the relativistic case, Eqs. (1.32), (1.33), (1.34), and (1.35), is called synchrotron radiation. The emission radiated by high-energy (cosmic-ray) electrons in the galactic disk interacting with the interstellar magnetic field is synchrotron radiation.

1.3 Thermal Emission

1.3.1 Thermal Bremsstrahlung

We have shown that an electron oscillating at a frequency ω emits radiation of intensity

$$I = \frac{2d^2\omega^4}{3c^3}. \tag{1.36}$$

The electron emits the radiation at the frequency $\omega = 2\pi\nu$ with ν the frequency of radiation. The emissivity by the particle between the frequency ν and $\nu + d\nu$ is given by

$$dI_\nu = Id\nu = \frac{2d^2 16\pi^4\nu^4}{3c^3}d\nu = \frac{32\pi^4 e^2 r_\nu^2 \nu^4}{3c^3}d\nu. \tag{1.37}$$

Suppose that the electron encounters an ion at an impact parameter ρ. Then the total radiation by an electron during the encounter can be expressed as

$$d\chi_\nu = \int dI_\nu 2\pi\rho d\rho = \text{const} \times \frac{e^4}{v^2c^3m^2}\ln\frac{2mv^3}{\gamma\omega e}d\omega, \tag{1.38}$$

where $\gamma = 1.78$.

In a plasma of temperature T, the free electrons have a Maxwell distribution in the velocity space:

$$f(v)dv = 4\pi\left(\frac{m}{2\pi kT}\right)^{3/2}\exp\left(-\frac{mv^2}{2kT}\right)v^2dv, \tag{1.39}$$

where k is the Boltzmann constant. The radiation from the plasma can be calculated by integrating the emissivity of individual electrons in the velocity:

$$\begin{aligned}
\varepsilon_\nu d\nu &= \int n_\mathrm{i}n_\mathrm{e} v d\chi_\nu f(v)dv \\
&= \text{const} \times \frac{e^6 n_\mathrm{i}n_\mathrm{e}}{\sqrt{kT}m^{3/2}c^3}\ln\frac{(2kt)^{3/2}}{\pi\gamma^{3/2}m^{1/2}\nu}d\nu \\
&= \frac{16e^6n_\mathrm{e}^2}{3\sqrt{2\pi}\sqrt{kT}m^{3/2}c^3}\ln\frac{0.21(kT)^{3/2}}{e^2\sqrt{m}\nu} \\
&= 3\times10^{-39}\frac{n_\mathrm{e}^2}{\sqrt{T}}\left(17.7 + \ln\frac{T^{3/2}}{\nu}\right).
\end{aligned} \tag{1.40}$$

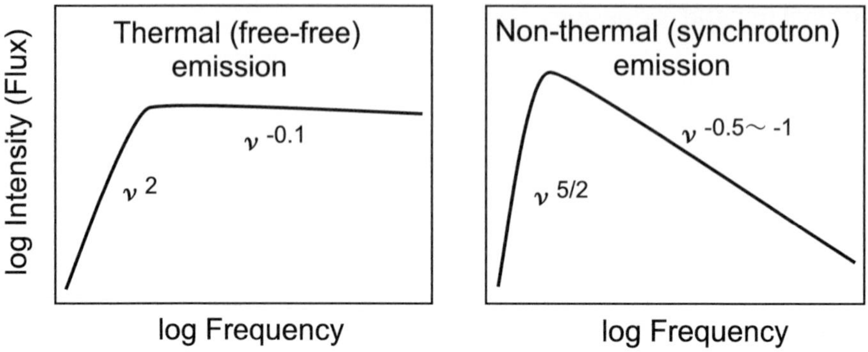

Fig. 1.1 Schematic spectra of thermal emission from ionized gas (*left*) and synchrotron emission from cosmic-ray electrons and magnetic field (*right*)

It is shown that in the radio frequencies with $h\nu \ll kT$ we have

$$\varepsilon_\nu \propto \nu^{-0.1} \tag{1.41}$$

However, if the emitting region is opaque, or the optical depth is large enough, the emission becomes close to a black-body radiation obeying Wean's law:

$$\varepsilon_\nu \propto \nu^2. \tag{1.42}$$

In Fig. 1.1 we illustrate the schematic spectrum of thermal radio emission.

1.3.2 Thermal Emission and Absorption Coefficient

The emissivity can be related to the Einstein coefficient as

$$\varepsilon_\nu d\nu = \frac{N_2}{4\pi} A_{21} h\nu. \tag{1.43}$$

The absorption coefficient can be related to the Einstein B coefficient as

$$\kappa_\nu d\nu = \frac{1}{4\pi}(N_1 B_{12} - N_2 B_{21}) h\nu. \tag{1.44}$$

In thermal equilibrium the A and B coefficients can be related to each other as

$$A_{21} = -\frac{2h\nu^3}{c^2} B_{21}. \tag{1.45}$$

We then have

$$\frac{\kappa_\nu}{\varepsilon_\nu} = \frac{\frac{N_2}{4\pi}B_{21}(\frac{h\nu}{kT})h\nu}{\frac{N_2}{4\pi}A_{21}h\nu} = \frac{B_{21}}{A_{21}}\frac{h\nu}{kT} = \frac{c^2}{2h\nu^3}\frac{h\nu}{kT} = \frac{c^2}{2\nu^2 kT}. \tag{1.46}$$

This yields the absorption coefficient

$$\kappa_\nu = \frac{c^2}{2\nu^2 kT}\varepsilon_\nu = 9.77 \times 10^{-3}\frac{n_e^2}{T3/2\nu^2}\left(17.7 + \ln\frac{T^{3/2}}{\nu},\right) \ (\mathrm{cm}^{-1}) \tag{1.47}$$

with T being measured in K, n_e in cm^{-3}, and ν in Hz. The optical depth is then defined by

$$\tau_\nu = \int \kappa_\nu dx \sim L\kappa_\nu \propto Ln_e^2. \tag{1.48}$$

Or more practically,

$$\tau_\nu \simeq 0.082T^{-1.35}\nu^{-2.1}EM, \tag{1.49}$$

where

$$EM = \int n_e^2 dx \sim Ln_e^2 \tag{1.50}$$

in pc cm^{-6} is the emission measure of the plasma, T and ν are measured in K and GHz, respectively, and L is the geometrical depth of the emission region.

1.4 Synchrotron Radiation: Nonthermal Radiation

1.4.1 Emissivity and Spectrum

A single relativistic electron of energy E spiraling around a magnetic field of strength B (Fig. 1.2) emits radio waves of intensity

$$I_E = \frac{2e^4B^2E^2}{m^2c^7} = c_3B^2E^2. \tag{1.51}$$

The energy distribution function of cosmic-ray electrons can be expressed as

$$f(E)dE = N(E)dE \propto E^{-\beta}dE, \tag{1.52}$$

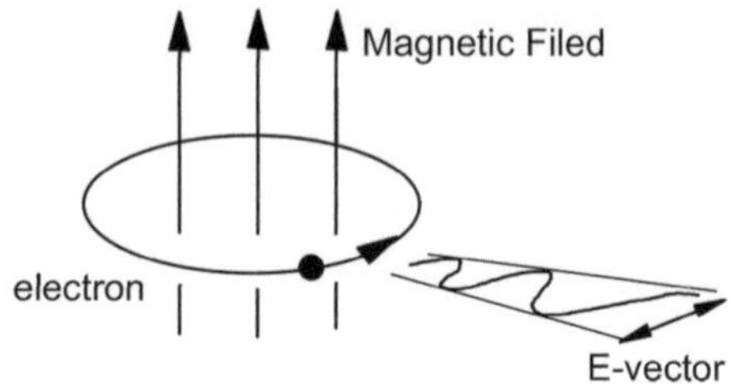

Fig. 1.2 Emission from an electron spiraling around magnetic lines of force

or

$$N(E)dE = N_0 \left(\frac{E}{E_0} \right)^{-\beta}. \tag{1.53}$$

For an optically thin case, the emissivity can be expressed as

$$dI_E = I_E f(E)dE = c_3 B^2 E^{2-\beta} dE. \tag{1.54}$$

Because the frequency of the emission is related to the energy through

$$\nu = \frac{eB}{mc}\gamma^2 = \frac{ecBE^2}{(mc^2)^3} \propto BE^2 \tag{1.55}$$

and

$$d\nu \propto BEdE, \tag{1.56}$$

we obtain the volume emissivity of the synchrotron radiation as

$$\varepsilon_\nu d\nu = \text{const} \times \nu^{\frac{2-\beta}{2}} \nu^{-1/2} d\nu = c_4 \nu^{\frac{1-\beta}{2}} d\nu. \tag{1.57}$$

It is known that observed cosmic-ray electrons have spectrum power with $\beta \sim 2 - 3$, which yields a spectral index of the synchrotron emission as $\varepsilon_\nu \propto \nu^{-0.5 \sim -1}$. On the other hand if the emission region is opaque (with sufficiently large optical depth), the radiation can be expressed by a black-body radiation with the temperature related to the energy as $kT = E$, and the emissivity expressed as

$$\varepsilon_\nu d\nu = E\nu^2 d\nu \propto \nu^{5/2} d\nu. \tag{1.58}$$

A schematic spectrum is shown in Fig. 1.1 in comparison with the spectrum of the thermal emission.

1.4.2 Energy Equipartition

If we adopt a simple assumption that the energy density (pressure) of the cosmic-ray electrons is in balance with the magnetic energy density (pressure), the synchrotron emissivity can be related to the magnetic field strength, averaged number density of high-energy electrons, and mean energy of the electrons. Hereby we assume that, under normal interstellar conditions, the synchrotron emission source is optically thin. The cosmic-ray electron energy density is given by

$$u_{\rm CR} = \int N(E) E dE \sim \langle NE \rangle, \tag{1.59}$$

and the magnetic energy density by

$$u_{\rm B} = \frac{B^2}{8\pi}. \tag{1.60}$$

The condition of the equipartition of the energy densities is expressed by

$$u_{\rm CR} \sim u_{\rm B}, \tag{1.61}$$

which yields

$$\frac{B^2}{8\pi} \sim \langle N \rangle \langle E \rangle \sim NE. \tag{1.62}$$

Let us also recall the expressions

$$\nu = \frac{ec}{(mc^2)^3} BE^2 \tag{1.63}$$

and

$$\varepsilon_\nu \sim \frac{e^4}{m^2 c^7} B^2 E^2 N \tag{1.64}$$

or the volume emissivity defined by

$$\varepsilon = L/V \sim \frac{e^4}{m^2 c^7} B^2 E^2 N. \tag{1.65}$$

Here L in erg s^{-1} cm^{-3} and V are the total luminosity and volume of the source, respectively. Equations (1.62), (1.63), (1.64), and (1.65) can then be used to obtain the following relations:

$$B \sim 3.29 \times 10^2 \nu_{\rm GHz}^{-1/7} \varepsilon^{2/7} \text{ (Gauss)}, \quad (1.66)$$

$$\langle N \rangle \sim 4.78 \times 10^9 \nu_{\rm GHz}^{-6/7} \varepsilon^{5/7} \text{ (cm}^{-3}\text{)}, \quad (1.67)$$

and

$$\langle E \rangle \sim 9.02 \times 10^{-7} \nu_{\rm GHz}^{4/7} \varepsilon^{-1/7} \text{ (erg)}, \quad (1.68)$$

where $\varepsilon = L/v$ is the emissivity measured in erg s^{-1} cm^{-3} and $\nu_{\rm GHz}$ is the frequency measured in GHz. The above equations can be used to obtain approximate quantities for B, $\langle N \rangle$, and $\langle N \rangle$ of cosmic-ray electrons using the observed quantity $\varepsilon = L/V$. The emissivity may also be roughly calculated by dividing the observed intensity by an assumed depth of the emitting region; $\varepsilon \sim \int \varepsilon_\nu d\nu \sim \nu I/L$.

1.5 Recombination Lines

The most frequently observed line spectra in astrophysics are the recombination line emission, molecular line emission, and the 21-cm neutral hydrogen (HI) line emission. We describe here the recombination line [1] (Fig. 1.3).

1.5.1 Frequency

The interstellar hydrogen gas is ionized by UV photons from massive stars of O and B types. Such ionized gas surrounding OB stars is called the HII region, which comprises ionized hydrogen gas of temperature 7,000–10,000 K. Free electrons emit

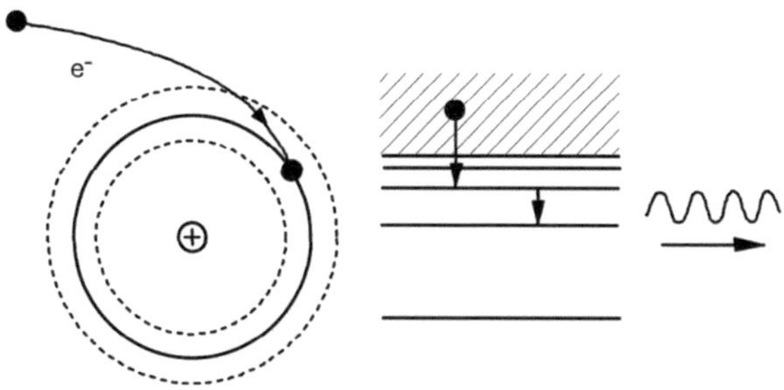

Fig. 1.3 Capture (recombination) of an electron by an H atom, and emission of recombination line

thermal Bremsstrahlung by free-free interaction with ions, and are recombined with the ions by emitting free-bound emissions. The recombined hydrogen is often at a high transition state, and emits bound-bound emission during its transition to the lower state. The emission is called the recombination line emission.

The energy difference between the upper and lower transition levels for a hydrogen atom is expressed by

$$h\nu = E_0 \left(\frac{1}{n_1^2} - \frac{1}{n_2^2} \right) \simeq 2E_0 \frac{\Delta n}{n^3}, \tag{1.69}$$

where $E_0 = h\nu_0 = hRcZ^2$ with R, Z, and n as the Rydberg constant, electric charge number, and the principal quantum number, respectively. For the hydrogen $Z = 1$ and $E_0 = 13.6\,\text{eV}$ ($\lambda_0 = 912\,\text{Å}$), $n_i \simeq n$ is the quantum number of the ith state, and $\Delta n (\ll n)$ is the level difference between the two states. We therefore obtain the frequency by

$$\nu \simeq 2RcZ^2 \frac{\Delta n}{n_1^3}. \tag{1.70}$$

For hydrogen, $Rc = 3.288051 \times 10^{15}$ Hz. The hydrogen recombination line for $\Delta n = 1$ is called the α line, and denoted the H$n\alpha$ line. For small n the line is observed in the optical wavelengths, whereas for larger n the lines are observed at the radio frequencies. In the microwave range we often observe the H$n\alpha$ line for n around 100; for example, the H110α line has the frequency of 4874.174 MHz.

1.5.2 Line Width

Inasmuch as line-emitting particles in a gas cloud of temperature T_k are moving because of thermal motion, the distribution function of the particles is expressed as

$$dN(v) = N \sqrt{\frac{M}{2\pi k T_\text{k}}} \exp\left(\frac{-mv^2}{2kT_\text{k}} \right) dv. \tag{1.71}$$

The Doppler effect of a line emission at ν_0 by the motion of a particle at velocity v is expressed as

$$\nu = \nu_0 \left(1 - \frac{v}{c} \right), \tag{1.72}$$

and the thermal velocity can be related to the frequency as

$$dv = -c \frac{d\nu}{\nu}. \tag{1.73}$$

The intensity of the line emission at ν is proportional to the number of particles at the velocity corresponding to this frequency,

$$dI(\nu) = If(\nu)d\nu = I\sqrt{\frac{4\ln 2}{\pi}}\frac{1}{\Delta\nu}\exp\left\{-4\ln 2\left(\frac{\nu-\nu_0}{\Delta\nu}\right)^2\right\}d\nu, \tag{1.74}$$

where $f(\nu)$ is the line shape function, and $\Delta\nu$ is the half- power width of the line:

$$\Delta\nu = \nu_0\sqrt{4\ln 2\frac{2kT_\mathrm{k}}{mc^2}}. \tag{1.75}$$

At $\nu = \nu_0$ we have $f(\nu_0) = 0.94\frac{1}{\Delta\nu} \simeq \frac{1}{\Delta\nu}$. For a hydrogen cloud, the observed line width is expressed as

$$\Delta\nu \simeq 7.1 \times 10^{-7}\nu_0\sqrt{T_\mathrm{k}} \tag{1.76}$$

with T_k being measured in K.

1.5.3 Line Intensity

Using the Einstein coefficients, we can write the emissivity and absorption coefficient of the line emission due to the transition from/to n_1 to/from n_2 generally as

$$\varepsilon_\nu d\nu = \frac{N_2}{4\pi}A_{21}h\nu \tag{1.77}$$

$$\kappa_\nu d\nu = \frac{1}{4\pi}(N_1B_{12} - N_2B_{21})h\nu. \tag{1.78}$$

For a thermal equilibrium, we have

$$N_1B_{12} = \exp\left\{\frac{h\nu}{kT}N_2B_{21}\right\}, A_{21} = \frac{2h\nu^3}{c^2}B_{21}. \tag{1.79}$$

For the recombination line, the absorption coefficient can be written as

$$\kappa_\nu = \frac{h\nu N_{n_1}}{4\pi\Delta\nu}\left\{1-\exp\left(\frac{-h\nu}{kT_\mathrm{e}}\right)\right\}B_{n_1n_2}. \tag{1.80}$$

We make use of the formula to give the density of partially ionized gas at n_1th level (Saha's formula)

$$N_{n_1} = \frac{N_e N_i h^3}{(2\pi m_e k T_e)^{3/2}} n_1^2 \exp\left\{\frac{-hcRZ^2}{n_1^2 k T_e}\right\}, \tag{1.81}$$

or

$$N_{n_1} \simeq 4.1 \times 10^{-16} n_1^2 \frac{N_e N_i}{T_e^{3/2}}. \tag{1.82}$$

Then we obtain

$$\kappa_\nu = \frac{\pi h e^2 N_{n_1} \nu}{mckT_e \Delta\nu} f_{n_1 n_2} = 1.04 \times 10^{-27} \frac{RcZ^2 \Delta n N_e Ni}{\Delta\nu T_e^{3/2}} \frac{f_{n_1 n_2}}{n_1}, \tag{1.83}$$

where

$$f_{n_1 n_2} = B_{n_1 n_2} \left(\frac{mch\nu}{4\pi^2 e^2}\right) \tag{1.84}$$

is called the oscillator strength. For $\Delta n = 1$ (Hα line) the oscillator strength has the value

$$\frac{f_{n_1 n_2}}{n_1} = 0.194, \tag{1.85}$$

and the above expression reduces to

$$\kappa_\nu = 6.7 \times 01^{-13} \frac{N_e N_i}{\Delta\nu T_e^{3/2}}, \tag{1.86}$$

and the optical depth is given by

$$\tau_{H\alpha} = \int_0^L \kappa\nu dl = 2.0 \times 10^6 \frac{\int_0^L N_e N_i dx}{\Delta\nu T_e^{3/2}} \simeq 2.0 \times 10^6 \frac{EM}{\Delta\nu T_e^{3/2}}. \tag{1.87}$$

The electron temperature T_e of HII regions is about 10^4 K, therefore the optical depth is expressed as

$$\tau_{H\alpha} \simeq 2.0 \times 10^{-4} \frac{EM}{\Delta\nu}. \tag{1.88}$$

Here, $EM = \int_0^L N_e N_i dl \sim N_e N_i L$ is the emission measure.

1.5.4 Line to Continuum Intensity Ratio and Temperature Determination

The brightness temperature T_B of a radio-emitting region is defined by the relation to the intensity by Planck's law

$$I_\nu = \frac{2h\nu^3}{c^2}\frac{1}{\exp(h\nu/kT_B) - 1} \simeq \frac{2kT\nu^2}{c^2} = \frac{2kT_B}{\lambda^2} \tag{1.89}$$

for the Rayleigh-Jeans approximation when $h\nu \ll kT$. If the optical depth is small enough, the brightness temperature is related to the electron temperature of the gas through

$$T_B = T_e(1 - e^{-\tau}). \tag{1.90}$$

The brightness temperature of the recombination line emission is given by

$$T_{\text{B: Line}} \simeq T_e(1 - e^{-\tau_{\text{H}\alpha}}) \simeq T_e\tau_{\text{H}\alpha} \simeq 2 \times 10^6 \frac{EM}{\Delta\nu T_e^{3/2}}. \tag{1.91}$$

Recalling that the line width $\Delta\nu$ is related to the frequency and temperature by

$$\Delta\nu = 7.1 \times 10^{-7}\nu\sqrt{T(\text{K})} \tag{1.92}$$

we obtain

$$T_{\text{B: Line}} \simeq 3 \times 10^3 \frac{EM(\text{cm}^{-6}\text{pc})}{\nu_{\text{GHz}} T_e(\text{K})^2}. \tag{1.93}$$

On the other hand, the brightness temperature due to the free-free transition emission (thermal Bremsstrahlung) can be written as

$$T_{\text{B:Conti}} \sim T_e\tau_\nu \simeq 0.03\, EM\, T_e^{-1/2}\nu_{\text{GHz}}^{-2.1}. \tag{1.94}$$

Here, EM is in units of cm^{-6} pc and T in K. Hence, the intensity ratio of the recombination line to the continuum can be obtained by

$$\frac{T_{\text{B:Line}}}{T_{\text{B:Conti}}} \simeq 1 \times 10^5 \nu_{\text{GHz}}^{1.1} T_e^{-3/2}. \tag{1.95}$$

This equation shows that the electron temperature of an HII region can be obtained simply from the line-to-continuum intensity ratio:

$$T_e(\text{K}) \simeq 2.1 \times 10^3 \nu_{\text{GHz}}^{0.73} \left(\frac{T_{\text{B:Conti}}}{T_{\text{B:Line}}}\right)^{2/3}. \tag{1.96}$$

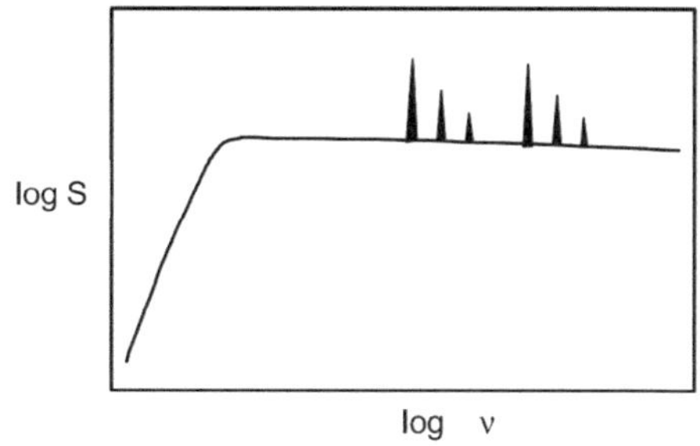

Fig. 1.4 Thermal radio continuum emission and the recombination lines

Figure 1.4 illustrates the spectrum of thermal radio emission superposed by recombination lines, where the line to the continuum intensity ratio can be measured.

1.6 Molecular Lines

Interstellar molecules comprising different kinds of elements, such as CO, are asymmetric with respect to their rotation axes. Molecules are asymmetrically charged due to friction during collisions with other molecules such as H_2. Such a molecule mimics a rotating charged particle that has an electric dipole moment, and emits radiation in the mm-wave range. The dipole moment for a CO molecule is $\mu = 0.112$ Debye $= 1.12 \times 10^{-19}$ (esu).

1.6.1 Frequency

The most abundant molecule in the interstellar space is carbon monoxide CO. A rotating molecule of CO, for example, emits dipole emission at its rotation frequency given by

$$\nu \sim \frac{v_{\rm r}}{2\pi a}, \tag{1.97}$$

where v and a are the rotation velocity of the molecule and its length (interval of C and O atoms) (Fig. 1.5). Because the rotation is excited by collisions with H_2 atoms in thermal motion, the rotation velocity can be approximately related to the thermal velocity v of H_2 atoms as

$$m_{\rm CO} v_{\rm r} \sim m_{\rm H_2} v, \tag{1.98}$$

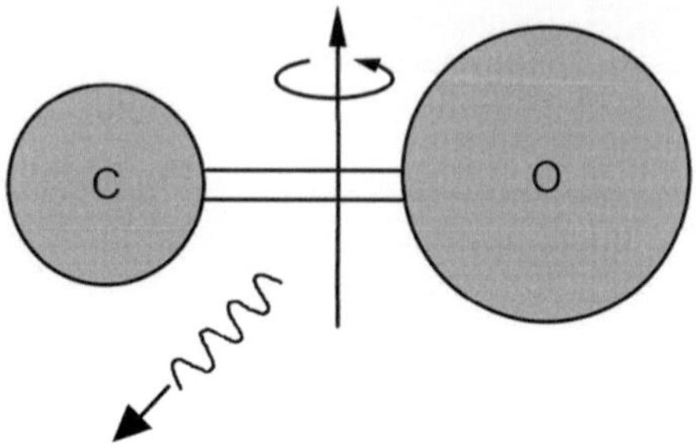

Fig. 1.5 CO molecule: the rotation and emission

where m_i are the masses of CO and H_2 molecules. We therefore have

$$v_r \sim v \frac{m_{H_2}}{m_{CO}} \sim 0.1v. \tag{1.99}$$

The thermal velocity v in a typical molecular cloud, approximately corresponding to the first-excited state of the CO molecule, is about $1\ \mathrm{km\,s^{-1}}$. The frequency of the dipole emission from this CO molecule with $a \sim 1.5$ A is then estimated to be

$$\nu \sim \frac{0.1\ \mathrm{km\ s^{-1}}}{2\pi a} \sim 1 \times 10^2\ \mathrm{GHz}. \tag{1.100}$$

The energy level for a rotation quantum number J of a CO molecule is given by

$$E_{\mathrm{rot}} = hBJ(J+1), \tag{1.101}$$

where h is the Planck constant, and B is related to the moment of inertia I as

$$B = \frac{h}{8\pi I}. \tag{1.102}$$

The frequency of a photon emitted by a transition from $J+1$ to the Jth rotational state is given by

$$h\nu = \Delta E_{\mathrm{rot}} = 2(J+1)hB. \tag{1.103}$$

The frequency of the rotational transition from $J = 1$ to $J = 0$ of a CO molecule is

$$\nu_{J=1-0} = 115.271204\ \ \mathrm{GHz}, \tag{1.104}$$

or $\lambda = 2.6$ mm.

1.6.2 Intensity

Recalling that the emission and absorption coefficients can be written by the Einstein coefficients as

$$\varepsilon_\nu d\nu = \frac{N_2}{4\pi} A_{21} h\nu, \tag{1.105}$$

$$\kappa_\nu d\nu = \frac{1}{4\pi}(N_1 B_{12} - N_2 B_{21}) h\nu, \tag{1.106}$$

we obtain the absorption coefficient for the molecular line radiation as

$$\begin{aligned}\kappa_\nu &= \frac{c^2}{8\pi\nu^2}\left(\frac{g_{J+1}}{g_J}N_J - N_{J+1}\right)A_{J+1}\frac{1}{\Delta\nu} \\ &= \frac{c^2}{8\pi\nu^2}\left\{1 - \exp\left(\frac{-h\nu}{kT_{J+1,J}}\right)\right\}\frac{g_{J+1}}{g_J}A_{J+1}\frac{1}{\Delta\nu}.\end{aligned} \tag{1.107}$$

Here, g are the degenerated level number, $g_J = 2J + 1$, and $g_{J+1} = 2J + 3$, and

$$\frac{N_{J+1}}{N_J} = \frac{g_{J+1}}{g_J}\exp\left\{\frac{-h\nu}{kT_{eJ+1,J}}\right\}. \tag{1.108}$$

The A coefficient can be related to the dipole moment as

$$A_{J+1,J} = \frac{64\pi^4}{3hc^3}|\mu_{J+1,J}|^2 = \frac{64\pi^4}{3hc^3}\mu^2\frac{J+1}{2J+3}. \tag{1.109}$$

Here, let us recall that the intensity of dipole emission from an electric dipole rotating at frequency ν is given by $I_\nu = 2\mu^2(2\pi\nu)^4/(3c^3)$.

If the molecule is in thermal equilibrium of temperature T, we have $T_{J+1,J} = T$, and the Boltzmann distribution of N_J can be expressed as

$$N_J = \frac{ng_J}{Q}\exp\left\{\frac{-hBJ(J+1)}{kT}\right\}, \tag{1.110}$$

where n is the number density of the molecule, and Q is the partition function;

$$Q = \sum_{J=0}^{\infty}(2J+1)\exp\left\{\frac{-J(J+1)hB}{kT}\right\} \simeq \int_0^\infty (2x+1)\exp\left\{-x(J+1)\frac{hB}{kT}\right\} = \frac{kT}{hB}. \tag{1.111}$$

When J is small enough, we finally obtain

$$\kappa_\nu \simeq \frac{4\pi^3 h}{3c(kT)^2} \frac{2J+3}{2J+1} \frac{\nu^3 \mu^2 n}{\Delta\nu}. \tag{1.112}$$

1.6.3 H_2 Mass Estimated from the CO Intensity

The brightness temperature is related to the temperature by $T_B = \tau T \simeq \kappa_\nu T L$ when $\tau \ll 1$, and is given by

$$T_B \simeq 9.3 \times 10^{-9} \frac{n_{CO} L}{\Delta\nu T}. \tag{1.113}$$

Using the Doppler relation we have

$$N_{CO} = \int n_{CO} dx \simeq n_{CO} L \tag{1.114}$$

$$\simeq 4.1 \times 10^{13} T \int T_B dv \simeq 4.1 \times 10^{13} T\, T_B \Delta v = 4.1 \times 10^{13} T\, I_{CO}. \tag{1.115}$$

Here, N_{CO} is the column density of CO molecules in cm^{-2}, and I_{CO} is called the CO line intensity. The temperature and velocity are measured in K and km s^{-1}, respectively. If the observed cloud is optically thin, the column density of H_2 molecules can be estimated by assuming the abundance of CO molecules, which is approximately $[CO]/[H_2] \sim 10^{-5}$ and yields

$$N_{H_2} \sim 4 \times 10^{18} I_{CO}. \tag{1.116}$$

Figure 1.6 illustrates a molecular cloud composed of turbulent cloudlets whose integrated line-width represents the velocity dispersion of the cloud. The size and the velocity dispersion are used in the Virial theorem to calculate the mass of the cloud.

However, interstellar molecular clouds are usually opaque against the CO line. An interstellar cloud comprises numerous turbulent eddies, each of which is also optically thick for the CO line. Because the eddies have different velocities and therefore different absorbing frequencies, the cloud can be seen through because of its velocity (frequency) gradient along the line of sight. This assumption is called the large-velocity-gradient (LVG) condition.

The mass of a molecular cloud can be estimated by the velocity gradient (velocity width) and its radius by the Virial relation. Observations have shown that the Virial mass (total mass) of a molecular cloud is proportional to its CO luminosity

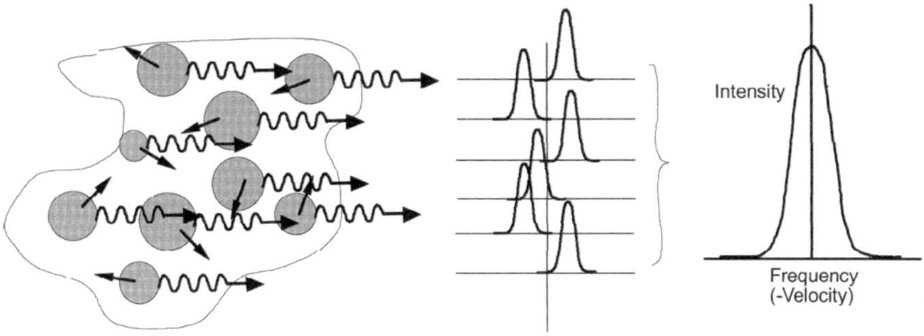

Fig. 1.6 Molecular cloud and its eddies, emitting CO line radiation

(integrated intensity over the extent). We have an empirical relation that relates the CO intensity to the molecular hydrogen column density as

$$N_{H_2} \sim 2.8 \times 10^{20} I_{CO}, \tag{1.117}$$

where N_{H_2} and I_{CO} are measured in $H_2\,cm^{-2}$ and $K\,km\,s^{-1}$, respectively.

1.6.4 Other Molecules

Numerous kinds of interstellar molecules have been observed thus far, such as ^{12}RCO (CO), ^{13}CO, H_2CO, HCN, $HC_{2n+1}N$ ($n = 0, 1, 2, \ldots 5$), and CS at various different transition states. They can be used not only to obtain the chemistry of the interstellar matter but also to obtain information about the kinematics (dynamics) and density distribution of a cloud. Each species of the molecules is representative of a certain interstellar condition, and has its characteristic distribution and kinematics in a cloud. The CO gas is distributed widely in any molecular clouds of various sizes, from core to envelope, and is most abundant; therefore, it is the most easily observed line among the interstellar molecular lines. Molecules such as H_2CO, HCN, and $HC_{2n+1}N$ are found in high-gas density regions, and therefore are used as probes to observe the cores and regions more directly related to the star formation in molecular clouds.

1.7 HI Line

The most abundant species of interstellar matter is molecular hydrogen (H_2) and neutral hydrogen (HI). The H_2 gas is usually found in low-temperature molecular clouds (~5–50 K)of sizes from approximately a few to 30 pc, and is distributed in

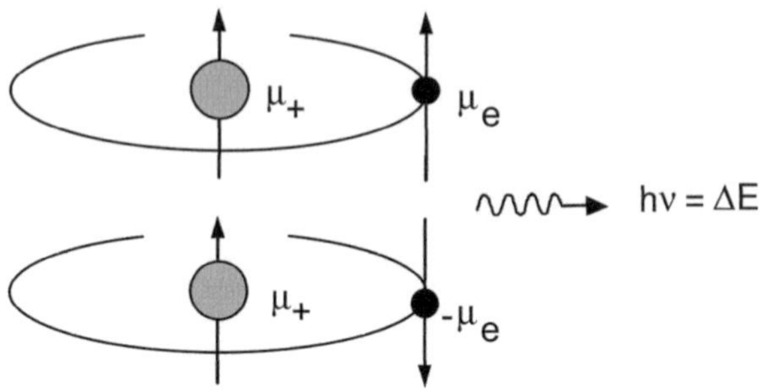

Fig. 1.7 An HI atom comprises two spins (magnetic dipoles) of the proton and electron, and their transition emits the HI 21-cm line radiation

the inner region of the galaxy at radius <10 kpc from the center. The HI gas is found in clouds of larger sizes, 10–100 pc, at temperatures of around 100 K, and is also distributed diffusely in the interstellar space and in the galaxy for a radius up to 20–30 kpc. The neutral hydrogen emits the 21-cm HI line.

1.7.1 Frequency

An H atom comprises a proton and an electron (Fig. 1.7). The proton has the positive charge $+e$ and is rotating (spinning) to yield a magnetic dipole moment $\mu_{\rm p}$. The electron is also rotating with the charge $-e$ to have the magnetic dipole moment $\mu_{\rm e}$. When the two magnetic dipoles are anti-parallel, the state is stable. If the spins are parallel, the atom emits line radiation at the 21 cm wavelength through the transition from the parallel to anti-parallel spin state. The parallel state is realized when the atom is excited such as because of collisions with other atoms. The energy of a photon emitted by this transition is given by

$$h\nu = \Delta E = \frac{\mu_{\rm e}\mu_{\rm p}}{a^3}, \tag{1.118}$$

where a is the Bohr radius:

$$a = \frac{h}{2\pi m_{\rm e} c\alpha}, \tag{1.119}$$

$$\mu_{\rm e} = \frac{eh}{4\pi m_{\rm e} c}, \tag{1.120}$$

$$\mu_{\rm p} = \frac{geh}{4\pi m_{\rm p} c}, \tag{1.121}$$

where $\alpha = 1/137$ is the fine-structure constant and g is the Landé factor of order unity.

The energy level can be expressed in terms of the principal quantum number n, the total angular momentum F of the atom, $F = J + I = L + S + I$, where L is the orbital angular momentum of the electron, I the nucleus spin angular momentum, S the electron spin angular momentum, and J is the total angular momentum of the electron, respectively;

$$E = \frac{h\nu_0}{n^3} \frac{F(F+1) - I(I+1) - J(J+1)}{J(J+1)(2L+1)}, \tag{1.122}$$

with $\nu_0 = g\alpha^2 cR(m_e/m_p)$ and $R = 2\pi^2 m_e e^4/(ch^3)$ is the Rydberg constant. The transition energy from $F = 1$ (parallel spin) to $F = 0$ (anti-parallel) state is then written as

$$\Delta E = E(F = 1) - E(F = 0). \tag{1.123}$$

The frequency of the HI line emission is given as

$$\nu = 1420.405751786 \text{ MHz}, \tag{1.124}$$

or

$$\lambda = 21.106114 \text{ cm}. \tag{1.125}$$

1.7.2 HI Intensity and Column Density

The absorption coefficient of the HI line radiation is expressed in terms of the Einstein A coefficient

$$\begin{aligned} \kappa_\nu &= \frac{c^2 A_{10}}{8\pi\nu^2} n_0 \frac{g_1}{g_0} \left\{ 1 - \exp\left(\frac{-h\nu_{10}}{kT}\right) \right\} f(\nu) \simeq \frac{c^2 A_{10}}{8\pi\nu_{10}} \frac{3hn_H}{4kT} f(\nu) \\ &\simeq 2.6 \times 10^{-15} \frac{n_H}{T} \frac{1}{\Delta\nu}. \end{aligned} \tag{1.126}$$

For an optically thin gas the brightness temperature and the optical depth can be related by

$$T_B = T(1 - e^{-\tau_\nu}) \simeq T\tau = T \int \kappa_\nu dx \simeq TL\kappa_\nu. \tag{1.127}$$

Fig. 1.8 HI line emission by a cloud and absorption of continuum source

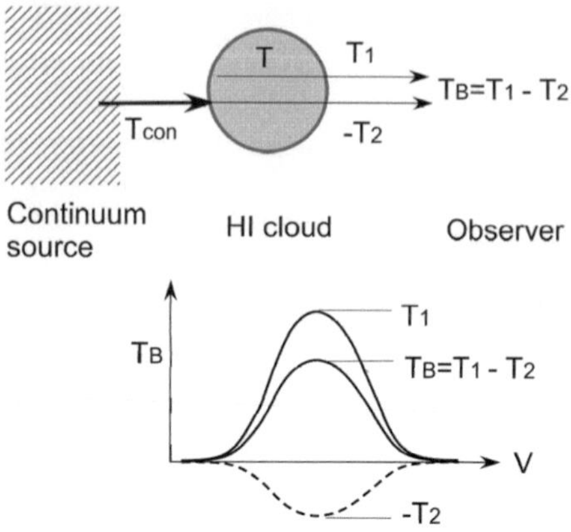

This equation is used to relate the observed brightness temperature T_B of the HI line and radial (Doppler) velocity v, which is a function of the frequency ν, to the column density of HI gas N as

$$N = \int n_H dx = 3.8 \times 10^{14} \int T_B dv = C_{HI} \int T_B dv, \tag{1.128}$$

where the column density, temperature, and velocity are measured in H cm^{-2}, K, and km s^{-1}, respectively, and $C_{HI} = 1.82 \times 10^{18}$ cm^{-2} (K km s^{-1-1}) is the conversion factor from HI intensity to column density.

In more general cases with finite optical depth as illustrated in Fig. 1.8 including thin as shown above, these expressions are written as

$$N_H = -C_{HI} T \int \ln\left(1 - \frac{T_B}{T}\right) dv, \tag{1.129}$$

with the excitation temperature T measured in K. If there is a radio continuum source behind the HI cloud, the cloud yields an absorption line against the continuum emission. This results in the following expression for the column density.

$$N = -C_{HI} T \int \ln\left(1 - \frac{T_B}{T - T_{con}}\right) dv, \tag{1.130}$$

where T_{con} is the brightness temperature of the radio continuum source. Figure 1.9 shows the ratio of H atom density calculated for finite optical thickness to that calculated under an assumption that the line is optically thin as a function of observed brightness temperature in the presence of a background continuum source. The excitation (spin) temperature of the HI gas is assumed to be 130 K.

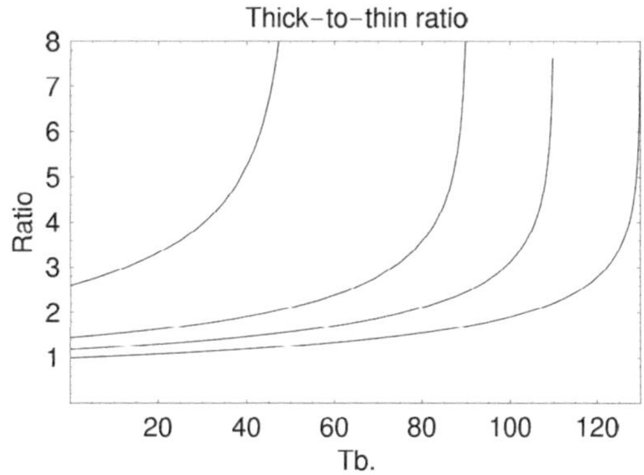

Fig. 1.9 Ratio of volume densities calculated for optically thick to thin cases for continuum emissions with $T_{\rm con} = 0$, 20, 40, and 80 K

1.7.3 HI Volume Density from Brightness Temperature

The volume density of an HI gas is given by

$$n = \frac{dN}{dx} = \frac{dN}{dv}\frac{dv}{dx} \tag{1.131}$$

and, hence, is calculated by observing the HI brightness temperature as

$$n = C_{\rm HI} T_{\rm B} \frac{dv}{dx} \tag{1.132}$$

for the optically thin case. In the general case n is given by

$$n = -C_{\rm HI} T \ln\left(1 - \frac{T_{\rm B}}{T}\right)\frac{dv}{dx}, \tag{1.133}$$

or

$$n = -C_{\rm HI} T \ln\left(1 - \frac{T_{\rm B}}{T - T_{\rm con}}\right)\frac{dv}{dx}, \tag{1.134}$$

when there is a continuum source behind the HI cloud.

The ratio of densities calculated for Eqs. (1.133) and (1.132) is given by

$$\frac{n_{\rm thick}}{n_{\rm thin}} = -\frac{T}{T_{\rm B}} \ln\left(1 - \frac{T_{\rm B}}{T - T_{\rm con}}\right) \tag{1.135}$$

The ratio $n_{\rm thick}/n_{\rm thin}$ calculated for excitation temperature $T = 130$ K at various observed brightness temperatures $T_{\rm B}$ is shown in Table 1.1.

Table 1.1 Ratio of HI column densities for optically thick to thin assumptions for HI excitation temperature $T = 130\,\mathrm{K}$

$T_B =$	10	20	50	100	120z	129	130 K
$n_{\mathrm{thick}}/n_{\mathrm{thin}} =$	1.04	1.09	1.26	1.91	2.78	4.91	∞

1.8 Radiations from Various Species

We summarize the kinds of radiations from various interstellar matter in the order of their temperature (energy) from low to high.

1.8.1 Molecular Lines

Molecular lines are emitted by the low-temperature (5–50 K) and high density ($10^3 \sim 10^6$ $H_2\,\mathrm{cm}^{-3}$) interstellar molecular clouds. Numerous species have been observed and can also be used to probe the chemistry of clouds. The lines can be used to probe the internal kinematics and density distribution of the individual clouds, particularly the regions around the star formation and dense galactic arms and rings. The molecular gas, mainly mapped in the CO line emission, is concentrated in the inner region of the galaxy within radius $\sim$10 kpc, particularly in the 4-kpc molecular ring and the nuclear disk of a few hundred pc radius.

1.8.2 HI Line Emission

The HI line is emitted by neutral hydrogen gas of temperature around $\sim$100–1000 K and density of $\sim$1–100 H cm^{-3}. The HI gas is distributed all over the galaxy, forming a broad ring of radius 10 kpc. The 21-cm line is used to obtain kinematic structure of the galaxy such as the galactic rotation curve. The velocity information along the galactic plane can also be used to obtain the face-on distribution of the interstellar gas, and of the spiral arms. The majority of the HI gas forms HI clouds of sizes 10–100 pc of density 10–100 dH cm^{-3}, and a part is diffusely distributed in the galactic disk of thickness 100–200 pc.

1.8.3 Recombination Lines

Recombination lines are emitted by partially ionized gas in HII regions of temperature $\sim 6000 - 10^4$ K and density $\sim 100 - 10^4$ H cm^{-3}. Spectroscopic data can be used to study kinematics of HII regions. The line intensity, combined with the thermal free-free transition emission, is used to obtain the density and temperature of the HII region. HII regions are associated with and excited by young massive stars (OB stars) that emit strong UV radiation.

1.8.4 Free-Free Emission

Free-free emission (thermal Bremsstrahlung) is a continuum radiation from plasma (ionized gas) of temperature $10^4 \sim 10^7$ K. The intensity is directly related to the emission measure $\int n_e^2 dx \sim n_e^2 L$ (cm^{-6} pc), and is used to obtain the ionized gas density of HII regions.

1.8.5 Synchrotron Radiation

The synchrotron (nonthermal) emission is a continuum radiation caused by the interaction of relativistic electrons (cosmic-ray electrons) with the interstellar magnetic field [3, 8, 9]. The volume emissivity $\sim I/L$ (ergs s^{-1} cm^{-3}) of the emission is used to measure the strength of the magnetic field and the cosmic-ray electron energy as well as the number density of high-energy electrons, under the assumption of the energy equipartition between magnetic and cosmic-ray pressures (energy densities). The radiation is usually polarized perpendicular to the magnetic lines of force.

This fact can be used to obtain information about the orientation of interstellar magnetic fields. The synchrotron radiation is emitted by the galactic disk, the nucleus (Sgr A), supernova remnants, pulsars, and the magnetic regions surrounding the galactic center. Radio galaxies, quasars, and usually spiral galaxies, are also strong sources of synchrotron radiation.

1.8.6 Black-Body Radiation

Finally, black body radiation is observed toward objects in which the optical depth is large enough (optically thick) (Fig. 1.10). For a nonrelativistic case, black-body radiation is expressed by Planck's law:

$$I_\nu(T)d\nu = \frac{2h\nu^3}{c^2} \frac{1}{\exp(h\nu/kT) - 1} d\nu \tag{1.136}$$

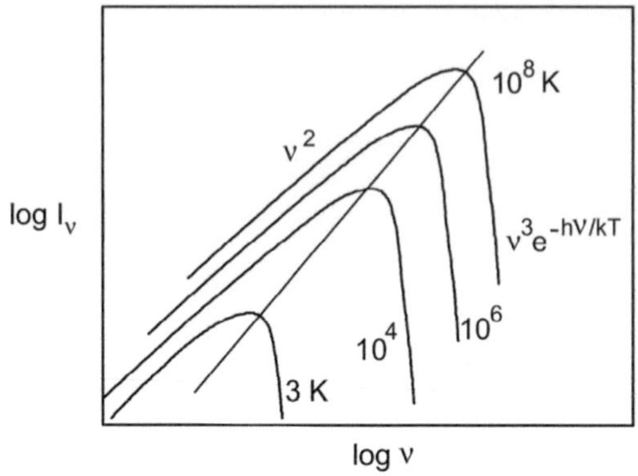

Fig. 1.10 Black-body radiation spectrum

in ergs s^{-1} cm^{-2} str^{-1}, or in terms of wavelengths,

$$I_\lambda(T)d\lambda = \frac{2hc^2}{\lambda^5}\frac{1}{\exp(hc/\lambda kT) - 1}d\lambda. \tag{1.137}$$

In the radio frequency range ($h\nu \ll kT$), the Rayleigh-Jeans law is applied;

$$I_\nu = \frac{2kT}{c^2}\nu^2 \propto \nu^2. \tag{1.138}$$

For a relativistic gas with an equivalent temperature of $kT \sim E$ with E being the energy of a high-energy particle, we obtain an optically thick synchrotron spectrum;

$$I_\nu \propto \nu^{5/2}. \tag{1.139}$$

1.9 Radiative Transfer

Radiation is absorbed during its propagation through interstellar matter (Fig. 1.11). If a source with intensity I lies behind an interstellar cloud of absorption coefficient κ, a small fraction of the intensity

$$dI = -\kappa I dx \tag{1.140}$$

is absorbed during its penetration for a small path length dx, or

$$\frac{dI}{dx} = -\kappa I. \tag{1.141}$$

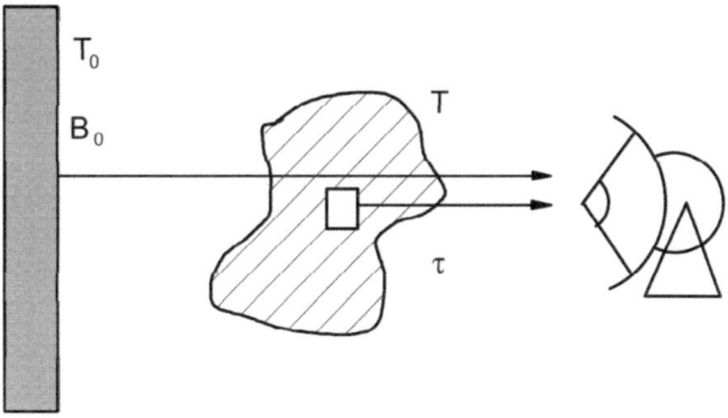

Fig. 1.11 Radiation transfer through a cloud

This is solved to give

$$I = I_0 e^{-\int \kappa dx}. \tag{1.142}$$

By introducing an optical depth

$$\tau = \int \kappa dx \tag{1.143}$$

we have

$$I = I_0 e^{-\tau}. \tag{1.144}$$

If there is a radio-emitting cloud with the emissivity ε on the path, we have the source term in addition to the absorption term in the above expression:

$$dI = -\kappa I + \varepsilon, \tag{1.145}$$

or

$$\frac{dI}{dx} + \kappa I = \varepsilon. \tag{1.146}$$

When κ =constant, the solution of this differential equation is

$$I = I_0 e^{-\int \kappa dx} + \frac{\varepsilon}{\kappa}(1 - e^{-\int \kappa dx}), \tag{1.147}$$

or

$$I = I_0 e^{-\tau} + \frac{\varepsilon}{\kappa}(1 - e^{-\tau}). \tag{1.148}$$

For an optically thin cloud, $\tau \ll 1$, we have

$$I = I_0 + \frac{\varepsilon}{\kappa}\tau. \tag{1.149}$$

If there is no background emitting source, $I_0 = 0$, we obtain

$$I = \frac{\varepsilon}{\kappa}\tau = S\tau. \tag{1.150}$$

The term

$$S = \frac{\varepsilon}{\kappa} \tag{1.151}$$

is called the source function. If the cloud is optically thick, $\tau \gg 1$, Eq. (1.148) yields

$$I = \frac{\varepsilon}{\kappa}, \tag{1.152}$$

or $I = S$. In solving Eq. (1.146) we assumed that κ is a constant. However, more generally, the absorption coefficient and the emissivity are functions of the depth along the line of sight. A differential equation of the type

$$\frac{dB}{dx} + f(x)B = g(x) \tag{1.153}$$

is called the Leibnitz equation. The general solution of this equation is given by

$$B = Ae^{-\int f(x)dx} + e^{-\int f(x)dx} \int e^{\int f(x)dx} g dx, \tag{1.154}$$

where A is a constant of integration.

For example, for a plasma emitting free-free thermal radiation a useful expression often used in radio astronomy is:

$$T_{\mathrm{B}} = T_{\mathrm{e}}(1 - e^{-\tau}) \tag{1.155}$$

or

$$T_{\mathrm{B}} \simeq T_{\mathrm{e}}\tau \ \text{ (for optically thin case, } \tau \ll 1), \tag{1.156}$$

and

$$T_{\mathrm{B}} \simeq T_{\mathrm{e}} \ \text{ (for optically thick case, } \tau \gg 1), \tag{1.157}$$

whereT_{B} and T_{e} are the (observed) brightness temperature and the electron temperature of the cloud, respectively.

1.10 Radio Astronomical Observables

We here summarize some terminologies used in radio astronomical observations and their definitions as well as the units [2, 4].

1.10.1 Observables at the Antenna

Flux density (flux): In radio astronomical observations, the radiation from a celestial object is detected by an antenna as the flux density S_ν at a frequency ν, which is measured as the energy (erg) penetrating through a unit area (cm^2) per unit time (s) per unit frequency interval (Hz). Thus the flux density (or flux) S is measured in units of erg cm^{-2} s^{-1} Hz^{-1}, and is often measured in the MKS unit system in w m^{-2} Hz^{-1}. Otten used the unit of flux density as Jy (Jansky), defined by 1 Jy = 10^{-26} w m^{-2} Hz^{-1}.

Intensity: The flux density is the row quantity received by the antenna. It represents the energy flowing through the place of the observer $I_\nu = S_\nu/\Omega_A$, which is called the intensity. Here, Ω_A is the solid angle of the observing beam of the antenna. The intensity is therefore measured in erg cm^{-2} s^{-1} Hz^{-1} str^{-1} or in w m^{-2} Hz^{-1} str^{-1}.

Surface brightness (brightness): When the source is extended in the sky having a solid angle Ω, the surface brightness of the source (or brightness) is defined as the flux density divided by the solid angle, $B = S/\Omega$, and is measured in erg cm^{-2} s^{-1} Hz^{-1} str^{-1} or in w m^{-2} Hz^{-1} str^{-1}. Note that the dimension is the same as the intensity, whereas the brightness is the quantity representing the specific intensity defined at the surface of the object and does not depend on the distance.

Brightness temperature: The surface brightness is often represented by an equivalent temperature at the measured frequency when the brightness is equaled by supposed brightness of a black body of temperature T_B usually measured in degree K. This temperature is called the brightness temperature, and is related to the surface brightness by $B = 2kT_B/\lambda^2$.

Antenna temperature: The row quantity measured by the antenna is often expressed by the antenna temperature T_A measured in degree K. It is defined as the temperature of a black body that supposedly surrounds the antenna system giving the same output voltage of the receiver as when the radio source is observed. This was originally introduced for convenience to calibrate the output of the receiver by inserting an absorber (a supposed black body) with known temperature (e.g., the room temperature) in front of the primary feed.

Main-beam temperature: Antennae have side lobes, thus the observed flux is not the total flux from the observed object. Hence, a correction for the lost flux by the side lobes is made by dividing the antenna temperature by the antenna efficiency η to yield the main-beam temperature, $T_{mb} = T_A/\eta$. Here, T_{mb} is used to calculate the flux density and is approximated to give the brightness temperature.

Frequency, wavelength, and velocity: Radio measurements are performed at fixed frequencies ν with finite bandwidths $\Delta\nu$. The frequency is related to the wavelength by $\lambda = c/\nu$. The frequency of line emission/absorption is related to the radial velocity v as $\lambda = \lambda_0(1 + v/c\sqrt{1-(v/c)^2})$, where λ_0 is the rest frequency at the source. In the interstellar and galactic radio astronomical observations, where $v \ll c$, the radial velocity is approximately $v \simeq \delta\lambda/\lambda \simeq -\delta\nu/\nu$.

1.10.2 Physical Observables After Conversion

Luminosity: The flux density can be converted to the radio emitting power of the source by multiplying the spherical surface area of the radius equal to its distance D as $L_\nu = 4\pi D^2 S_\nu$ and measured in erg s^{-1} Hz^{-}. The (total) luminosity of the source L is defined by $L = \int L_\nu d\nu$ and is measured in erg s^{-1}. If the distribution of the surface brightness is observed, the total luminosity of the source is given by $L = \int\int B_\nu d\nu d\Omega$.

Column density, or surface density: It often happens that we observe the line emissions such as the HI and CO lines. The line intensity is converted to the column density of gas by $N = C\int I_\nu d\nu$, where C is a conversion factor specifically determined to the emission line. In the case of interstellar gas the column density N is measured by the number of the gas particle in cm^{-2}.

Emission measure: If the line is due to collision-stimulated emission such as the thermal (free-free) emission, the intensity is related to the square of the emitting particles such as $EM = \int n_i n_e ds = C\int I_\nu d\nu$.

References

1. Brown, R.L., Lockman, F.J., Knapp, G.R.: Radio recombination lines. ARAA **16**, 445 (1978). [Recombination lines]
2. Cox, A. (ed).: Allen's Astrophysical Quantities. Springer, New York (1999). [General astrophysical quantities]
3. Ginzburg, V.L. Syrovatsikii, S.I.: Cosmic Magnetobremsstrahlung (synchrotron radiation). ARAA **3**, 297 (1965). [Synchrotron radiation]
4. Kraus, J.D.: Radio Astronomy. Cygnus-Quasar Books, Powell (1986). [Introduction to radaio astronomy]
5. Landau, L.D., Lifshitz, E.M.: Classical Theory of Field. Pergamon Press, New York (1960). [Emission mechanisms]
6. Pacholzyk, A.G.: Radio Astrophysics. Pergamon Press (1977). [Emission mechanisms]
7. Ribicke, G.B., Lightnman, A.P.: Radiative Processes in Astrophysics. Wiley, Weinheim (2004). [Radiation from plasma]
8. Schklovsky, I.S.: Cosmic Radio Waves. Harvard University Press, Cambridge (1960). [Synchrotron emission]
9. Syrovatskii, S.I.: Pinch sheets and reconnection in astrophysics. ARAA **19**, 163 (1981). [Synchrotron emission]

Chapter 2
Interstellar Matter

2.1 Energy Balance in ISM

Various kinds of gases and energetic particles composing the interstellar matter (ISM) are in hydrodynamical equilibrium when there exist no particular active regions such as star-forming regions and/or supernova explosions, and are embedded in the radiation field of stars [6, 10, 11] (Fig. 2.1). The gases radiate various kinds of radio waves according to their temperature and chemical compositions in continuum as well as in line emissions (Table 2.1). The population is the basic concept classifying interstellar matter in the galaxy, and the ISM is categorized as a Population I object (Table 2.2).

2.1.1 Energy-Density and Pressure Balance

Although ionized gas, dust, and cosmic rays are negligible concerning their mass, their energetic contribution is comparable to the other more massive species such as molecular clouds and HI gas. Let us estimate the energy density and the pressure of each species.

Energy density, which has the same dimension as the pressure, of molecular gas is given approximately as

$$u_{\rm mol} \sim n_{\rm H_2} kT \sim 10^{-12}\,{\rm erg\,cm^{-3}}, \tag{2.1}$$

with $n_{\rm H_2} \sim 10^3\ {\rm H_2\,cm^{-3}}$ and $T \sim 10$–$100\,{\rm K}$. The HI gas (clouds) has the amount of

$$u_{\rm HI} \sim n_{\rm H} kT \sim 10^{-12}\,{\rm erg\,cm^{-3}}, \tag{2.2}$$

The original version of this chapter was published without printing chapter abstract. The abstracts for each chapter are updated in the front matter in the current version of the book. An erratum to this chapter can be found at https://doi.org/10.1007/978-981-10-3445-9_7.

Y. Sofue, *Galactic Radio Astronomy*, Lecture Notes in Physics 935,
DOI 10.1007/978-981-10-3445-9_2

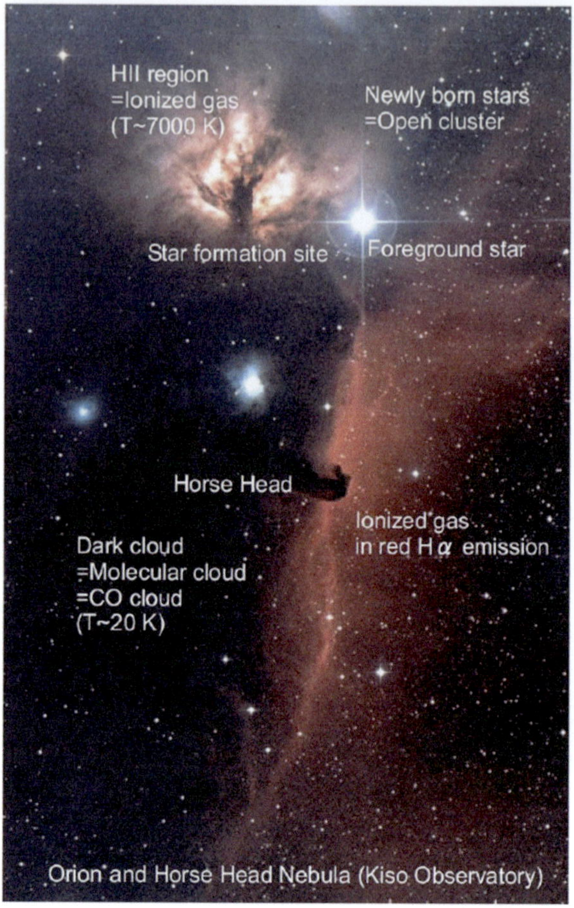

Fig. 2.1 The Orion and Horsehead Nebulae. They are composed of ionized gas emitting the red H_α line ($\sim$7000 K) in the HII region and of low-temperature ($\sim$20 K) molecular gas seen as the dark cloud silhouetted against background emission and stars (Courtesy: Kiso Observatory, The University of Tokyo)

with $n_{\rm HI} \sim 10\ {\rm H\,cm^{-3}}$ and $T \sim 100$–$1000\,{\rm K}$. The diffuse ionized gas in the intercloud regions has the energy density of

$$u_{\rm Diff.HII} \sim n_{\rm H} kT \sim 10^{-12}\,{\rm erg\,cm^{-3}}, \tag{2.3}$$

for $n_{\rm HII} \sim 10^{-2}\ {\rm H\,cm^{-3}}$ and $T \sim 10^4\,{\rm K}$. Therefore, the thermal components have the energy density of approximately $10^{-12}\,{\rm erg\,cm^{-3}} \sim 1\,{\rm eV\,cm^{-3}}$, or

$$u_{\rm gas} \sim 10^{-12}\,{\rm erg\,cm^{-3}} \sim 1\,{\rm eV\,cm^{-3}}. \tag{2.4}$$

Here, n_i is the number density of the ith gas species. These energy densities are also approximately equal to the kinetic energy:

Table 2.1 ISM and radiations

T(K), E	Object	Line	Continuum
2.7 K	CBR		mm,submm
10–50 K	Mol. cloud	Mol. lines	
	H_2 gas	mm,submm	
	Dust		FIR
100–1000 K	HI clouds	21-cm	
	Diffuse HI		
	Dust		IR
10^4 K	HII regions	Recom.lines	Free-free
$10^{5\sim7}$ K	SNR		Synch.
	Diffuse HII		X-rays
	Halo		
High-energy	Mag. fi, CR		Synch.Polari.
(keV-GeV)	SNR, Pulsars		
	Gal. Center		
	AGN, Jets		
	Radio gal.		
γ rays (GeV-TeV)	CR		γ rays

$$u_{\rm gas:kin} \sim \frac{1}{2}\rho_{\rm gas} v_\sigma \sim 10^{-12}\,{\rm erg\,cm^{-3}}, \tag{2.5}$$

where $\rho \sim$ 1–10 H cm^{-3} and $v_\sigma \sim$ 5–10 km s^{-1} are the gas density and velocity dispersion, respectively.

The high-energy components, cosmic rays, and magnetic fields, have energy densities as follows. The magnetic field energy density is

$$u_{\rm mag} = \frac{B^2}{8\pi} \sim (3 \times 10^6{\rm G})^2/8\pi \sim 10^{-12}\,{\rm erg\,cm^{-3}}. \tag{2.6}$$

This is in balance with the cosmic-ray energy density which is

$$u_{\rm CR} \sim 10^{-12}\,{\rm erg\,cm^{-3}}. \tag{2.7}$$

It must also be noted that the energy density of the starlight in interstellar space can be estimated as

$$u_{\rm starlight} \sim \frac{L}{\pi R^2}\frac{1}{c} \sim 10^{-12}\,{\rm erg\,cm^{-3}}, \tag{2.8}$$

where $L \sim 10^{11} L_\odot \sim 10^{44}$ ergs s^{-1} is the luminosity of the galaxy, and $R \sim 10$ kpc is the typical radius of the galactic disk. By the way, the cosmic background radiation of 3K has about the same amount of energy density:

Table 2.2 Population types of ISM and stars

	Pop. I Extr.Pop I young systems	Older pop I	Pop. II Old disk old systems	Intermed.	Halo
Gal. str.	Young s. disk Gas disk Bright arms		Old st. disk		Halo, Glo. cl.
Thickness (pc)	$z \sim 100$	200	300–500	1 kpc	2–10 kpc
Velo. (km s^{-1})	$v_\sigma \sim 5$–10		50–100		100–200
Mass($10^{11}\,M_\odot$)	0.05	0.05	1	1	~10
Age (Gy)	<0.1	0.1–1.5	1.5–5	5–6	6–15
Metallicity	$[Z/H] = 0.04$	0.02	0.01	0.004	0.001
ISM (Gas)	H_2 Dust Inner disk HI Diffuse HII		Outer HI		
Nebulae	Dark clouds Emiss. neb. HII regions SNR Type II Cirrus		Pla. neb. SNR Type		
Stars	Giants Metal rich T Tau Cepheids	Sun Nearby st. Main seq. Lines	Metal poor RR Lyr	High vel. weak lines	Glob. cl. Extr. poor RR Lyr

$$u_{3K} \sim \frac{\sigma T^4}{c} \sim 10^{-13}\,\mathrm{erg\,cm^{-3}}, \tag{2.9}$$

where $\sigma = 5.670 \times 10^{-5}\,\mathrm{erg\,cm^{-2}\,deg^{-4}\,s^{-1}}$ is the Stephan-Boltzmann constant and $T = 2.7\,\mathrm{K}$.

Thus, we have a general balance in the interstellar matter as

$$u_{\mathrm{gas}} \sim u_{\mathrm{mag}} \sim u_{\mathrm{CR}} \sim u_{\mathrm{starlight}} \sim u_{3K} \sim 1\,\mathrm{eV\,cm^{-3}}. \tag{2.10}$$

Because the energy density of a gas is equivalent to the pressure, we may say that the interstellar gases of various phases, magnetic fields, cosmic rays, and starlight pressure (on dust) are in a pressure balance with each other (Figs. 2.2 and 2.3).

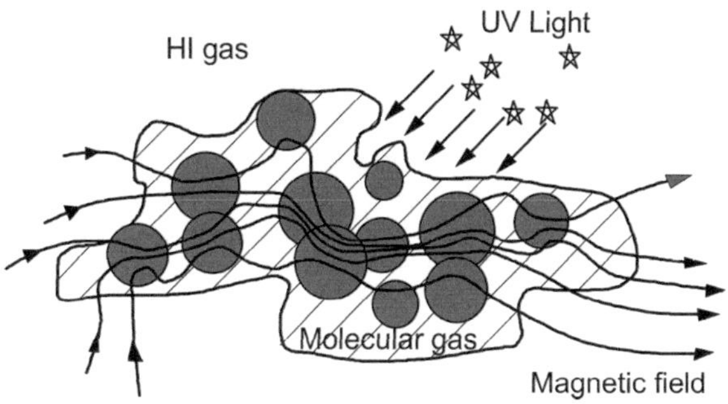

Fig. 2.2 The pressure balance (energy equipartition) in the ISM for an equilibrium and steady state

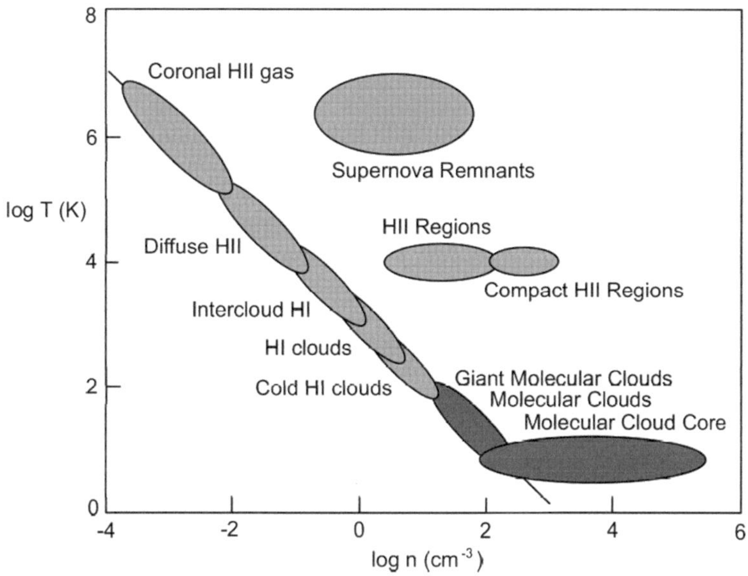

Fig. 2.3 Most of the gaseous objects lie on the line of pressure constant. Objects that are significantly displaced from the balance line are in an "activity"

2.1.2 "Activity" in ISM

Gaseous objects such as molecular clouds and HI clouds that compose the ISM are generally in thermal equilibrium and are balanced with each other by the mutual pressure. If the energy/pressure balance works, the clouds are stable and can exist in a steady state.

However, such a balanced (steady) state is often broken by an injection of extra energy. A supernova explosion results in a rapidly expanding shock-wave shell, and is observed as a supernova remnant (SNR). A sudden release of energy in the nuclei of galaxies often produces jets and fast outflow of gas as well as expanding shells. Strong UV radiation from newly born stars heats the ISM and leads to expansion of ionized gas as an expanding HII region.

The balance is also broken by the radiative cooling of gas by thermal instability followed by gravitational collapse of condensation to dense cores and sometimes leading to star formation. Supersonic collisions of clouds would also act to break the balance by sudden compression of gas and formation of new stars.

If an object is in such a nonequilibrium condition that the energy/pressure balance is broken, we may regard the object as being in an "activity." Active regions in the galaxy are the nucleus and its surroundings (galactic center region), star-forming regions, jets, supernovae and supernova remnants, inflating magnetic fluxes over the galactic disk, and so on.

2.2 Molecular Clouds

2.2.1 *Mass, Size, and Intensity*

Molecular gas is observed by spectral line observations in the mm-wave range, especially with the use of the CO molecular line (Fig. 2.4). The Virial (dynamical) mass of a molecular cloud of radius r and velocity dispersion v_σ is given by

$$M_{\rm cloud} \sim \frac{r v_\sigma^2}{G}. \tag{2.11}$$

The CO luminosity $L_{\rm CO}$ is related to the CO intensity $I_{\rm CO}$ as

$$L_{\rm CO} = 4\pi D^2 \int I_{\rm CO} d\Omega, \tag{2.12}$$

where D is the distance to the cloud and is related to r as $r \sim D\theta$ with θ the angular radius of the cloud. The distance can be obtained from the radial velocity of the cloud by using the galactic rotation law for an assumed rotation curve of the galaxy. The velocity dispersion v_σ can be obtained from the line width of the cloud's CO line profile.

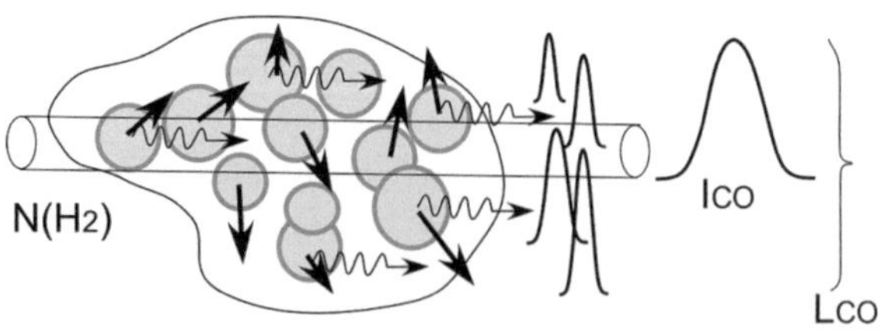

Fig. 2.4 Molecular cloud mass and intensity

The mass can be related to the column mass as

$$M_{\rm cloud} = 2m_{\rm H} \int N_{\rm H_2} dxdy = 2m_{\rm H} \int N_{\rm H_2} D^2 d\Omega. \tag{2.13}$$

From observations of many molecular clouds we have an empirical relation

$$M_{\rm cloud} \simeq A L_{\rm CO} \tag{2.14}$$

with A being a constant, although the CO line is optically thick. This gives us an empirical relation between the CO intensity and the column mass as

$$N_{\rm H_2} \simeq C I_{\rm CO} = C \int T_{\rm mb} dv, \tag{2.15}$$

where C is called the conversion factor, and has the value of

$$C = 2.0 \times 10^{20}\ ({\rm H_2\,cm^{-2}/K\,km\,s^{-1}}). \tag{2.16}$$

Here, $T_{\rm mb}$ is called the main-beam temperature when we observe it, and is approximately equal to the brightness temperature of the cloud. The conversion factor C is, of course, not a universal constant at all, and a value of $C = 3.6 \times 10^{20}$ ($\rm H_2\,cm^{-2}/K\,km\,s^{-1}$) is used sometimes. Although there is no definite observational evidence, the conversion factor must be larger in the central region of galaxies, where the metal abundance is higher than in the outer disk.

The factor also changes from galaxy to galaxy; in a metal-poor galaxy such as the Large Magellanic Cloud the value is about five times larger. A higher value has also been observed in the central 2 kpc region of M31, where some inflow of metal-poor gas from the companion may have occurred. Regarding the uncertainty, however, we assume the value quoted above for various circumstances in the galaxy and in extra galaxies when we need to obtain the mass of molecular gas.

The mass distribution of molecular clouds is known to have a form

$$f(M)dM \propto M^{-1.5}dM \tag{2.17}$$

with a lower and upper cutoff of about $\sim 10^2$ and $\sim 10^6\ M_\odot$. Clouds of mass around 10^3–10^5 are called normal molecular clouds, whereas those with 10^5–10^6 are called giant molecular clouds (GMC). Sizes of molecular clouds are 10–30 pc, and GMC 30–50 pc. The mean density of clouds is typically 10^3–10^4 $\rm H_2\,cm^{-3}$. The density of cores in clouds is sometimes 10^5–10^6 $\rm H_2\,cm^{-3}$, from which stars may form.

2.2.2 Distribution of Molecular Clouds

In usual spiral galaxies of Sb and Sc galaxies including the Milky Way, the density distribution of ISM (HI and H_2) is approximated by an exponentially decreasing function of scale radius of about 3–4 kpc [12].

Among various species of gases, molecular clouds are concentrated in the central region, making up the dense "nuclear gas disk." They again have a density peak at a galactocentric distance of 4 kpc, forming the "4-kpc molecular ring." They, particularly GMCs, also show enhancing along the spiral arms, and the major inner arms can be traced as well as in the HI line emission.

2.2.3 Giant Molecular Clouds

The largest (most massive) molecular clouds observed in the solar neighborhood have an extent of about $\sim$50 pc and velocity dispersion $v_\sigma \sim 5 - 10\ 5\,\mathrm{km\,s^{-1}}$. The Virial (dynamical) mass of a cloud can be estimated as $M \sim rv_\sigma^2/G \sim 1.6 - 6 \times 10^5\,M_\odot$ with $r \sim 1/2\times$ extent. Molecular clouds with a mass of several $10^5 \sim 10^6\,M_\odot$ are called giant molecular clouds.

GMCs are the most massive objects in the galaxy that are supposed to be gravitationally bound. In the galaxy, several 10^3 GMCs ($\sim$ a few $10^9\,M_\odot$ in total mass) are distributed in the thin galactic disk within a galactocentric distance of $R \sim 10$ kpc. They are further concentrated in the nuclear disk, the 4-kpc ring, and in the major inner ($R < 10$ kpc) spiral arms. They compose the molecular gas disk of a scale thickness of about $z \sim 60$ pc near the galactic plane.

The rest of the molecular gas (totaling $\sim 10^9\,M_\odot$) is in the form of normal molecular clouds (MC) which are less massive than GMCs. They are more diffusely distributed than GMCs, and are also distributed in the interarm regions.

Molecular gas is also sometimes observed at a height of several 100 pc above the galactic plane, often called high z molecular clouds. In some edge-on galaxies such as NGC 891, molecular clouds are distributed with a scale thickness of about 1 kpc. The distribution of MC in the galaxy away from the galactic plane (or in the halo) is still an open question, although a significant amount of MC is supposed to make up a thick molecular disk, as in NGC 891.

2.2.4 GMC and Star-Forming Sites

Stars are born through gravitational contraction of dense and cool interstellar gas. Cores of molecular clouds, particularly GMC, are, therefore, the places of star formation. GMCs are often associated with young massive stars (OB stars) and HII regions that developed due to UV radiation from these high-temperature stars.

Typical examples of the association of a star-forming region (OB star cluster and HII region) and a GMC are found in Ori GMC, the Ori Great Nebula, and the like. (Table 2.3).

Table 2.3 Molecular clouds and SF regions

Cloud	SF region	RA	Dec	l	b
Tau MC (cold)		04h 38.6m	$+25°36'$	172°	−13°
Ori GMC	Ori Neb, M42	05 h 32.8 m	−05 25	209	−19
Sgr B2 GMC	HII+Comp. HII	17 44.2	−28 23	1	−0
M17 GMC	SF HII	18 17.5	−16 15	15	−1
M16 GMC	HII & shell	18 17.5	−16 15	15	+0
Mon GMC	Ros. N & Mon. Lp	6 29.3	+04 57	206	−2
W49 MC	W49A HII	19 07.9	+09 01	43	−0
G24 MC + HI Ring	G24 HII Ring			24 30	+0

2.3 The CO-to-H_2 Conversion

Deriving the mass of molecular hydrogen gas is the fundamental process for the interstellar physics of galaxies [1, 7]. The principal method has adopted conversion from the intensity (or luminosity) of the $^{12}CO(J = 1 - 0)$ molecular line emission (I_{CO} or L_{CO}) into the column density (or mass) of H_2 molecules (N_{H_2} or M_{H_2}).

The conversion factor $C(= N_{H_2}/I_{CO} = M_{H_2}/L_{CO}) = C^* \times 10^{20}$ H_2/K km s^{-1} has been derived for molecular clouds in the solar vicinity on the assumption of a Virial equilibrium of individual clouds and the large-velocity-gradient for the CO line. This method has been used for deriving the conversion factor in nearby external galaxies such as M31, M33, and the Magellanic Clouds, where individual clouds can be resolved. A more global value in our galaxy has been derived by using a correlation between the distribution of γ-ray intensity, which is assumed to be proportional to the column mass of hydrogen atoms and molecules, and that of CO intensity along the galactic plane. A correlation between the CO intensity and optical extinction (A_v), which is assumed to be correlated with the column mass of hydrogen, has also been used, and has been applied to well-studied galaxies including M51.

The conversion factor obtained for the Milky Way Galaxy appears to be well defined around $C^* \sim 3$ within a factor of 1.3. However, the value seems to be largely scattered among galaxies on the order of magnitude, from our galaxy and galaxies of "normal metallicity" to such low-metallicity galaxies as the Magellanic Clouds ($C^* \sim 15$). Hence, it is a natural consequence of these facts that one supposes a dependency of C on metallicity. It is also natural to consider that the conversion factor is a function of the distance from the galaxy centers, because the metallicity exhibits a general exponential decrease with the galactocentric distance.

2.3.1 Metallicity Dependence of the Conversion Factor

The conversion factor varies with the metallicity of the interstellar gas. Figure 2.5 plots the conversion factor C^* against the oxygen abundance $12 + \log O/H$ for

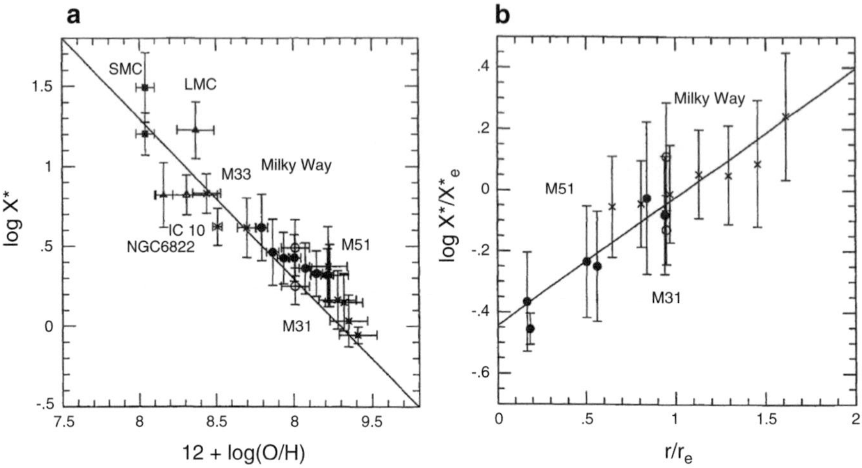

Fig. 2.5 (**a**) Conversion factor (C^*) versus oxygen abundance relation for eight disk galaxies. Different symbols indicate different galaxies. The *solid line* gives a least-square fit. (**b**) Radial distribution of conversion factor (C^*) for the Milky Way Galaxy, M31, and M51. The *solid line* gives the least-square fit

various galaxies for which metal abundance and the factor are known (e.g., M51, M31, the Milky Way Galaxy, M33, IC10, NGC6822, LMC, and SMC). The diagram demonstrates that the conversion factor depends on the metallicity of gas clouds. A simple least-squares fit gives

$$\log C^* = -0.80(12 + \log \mathrm{O/H}) + 7.60 \tag{2.18}$$

or equivalently,

$$C \propto Z(\mathrm{O})^{-0.8}, \tag{2.19}$$

where $Z(\mathrm{O}) = [\mathrm{O/H}]$ is the oxygen abundance.

This plot suggests that, among various properties of interstellar media such as the gas density and excitation temperature, the metallicity is a fundamental parameter that controls the CO-to-H_2 conversion factor. It is reasonable to suppose that the conversion factor is somehow dependent on the CO abundance $Z(\mathrm{CO})$. If we assume that O and C abundance are proportional to each other, the above equations indicate that the conversion factor is mainly controlled by $Z(\mathrm{CO})$ as

$$C \propto Z(\mathrm{CO})^{-0.8}. \tag{2.20}$$

2.3.2 Radial Variation of the Conversion Factor

The Milky Way Galaxy and M51 exhibit conspicuous radial gradients of the conversion factor. In Fig. 2.5, we plot the values of C^* for these galaxies as well as M31 as a function of a radius r/r_e, where r_e is an effective radius obtained from the stellar surface brightness distribution. The figure suggests that within uncertainties associated with data acquisition the $\log C^*$ versus r/r_e relation holds universally for the Milky Way Galaxy, M31, and M51, although r_e of each galaxy is not the same: $r_e = 6.19\,\text{kpc}$ for the Milky Way Galaxy, $r_e = 7.37\,\text{kpc}$ for M31, and $r_e = 7.26\,\text{kpc}$. A least-squares fit gives:

$$\log C^* = 0.37 + 0.40(r/r_e - 1). \tag{2.21}$$

Inasmuch as the radial distributions of C^* in these galaxies merely reflect the radial abundance gradient of oxygen, it is reasonable to have such a tight correlation between C^* and r/r_e. Indeed the oxygen abundance at any galactocentric distance in the disk of a spiral galaxy can be written as

$$(\text{O/H}) = (\text{O/H})_\text{e} 10^{-0.39(r/r_e - 1)}, \tag{2.22}$$

where $(\text{O/H})_\text{e}$ is the oxygen abundance at the effective radius of a galaxy. By putting Eq. (2.18) into Eq. (2.22), we obtain:

$$\log C^* = \log C_e^* + 0.31(r/r_e - 1), \tag{2.23}$$

where C_e^* is the conversion factor at the effective radius r_e. Equation (2.23) gives a nearly identical slope to that given by Eq. (2.21). This formula requires knowledge of C_e^* alone and can be used for any galaxy to which the oxygen abundance gradient is not actually known.

2.4 HI Gas and Clouds

2.4.1 Mass, Size, and Intensity of HI Clouds

For an optically thin case ($\tau \ll 1$), the column density of HI gas is estimated from the observed intensity of the 21-cm line emission as

$$N_\text{H}\,(\text{H}\,\text{cm}^{-2}) = 1.82 \times 10^{18} \int T_\text{B} dv = 1.82 \times 10^{18} I_\text{HI}\,(\text{K}\,\text{km}\,\text{s}^{-1}). \tag{2.24}$$

If the optical depth is close to unity or greater ($T_\text{B} \sim T_\text{ex}$) we have

$$T_\text{B} = T_\text{ex}(1 - e^{-\tau}), \tag{2.25}$$

with

$$\tau = N_{\rm H} 1.6 \times 10^{-15} \frac{1}{\Delta v T_{\rm ex}} = 5.5 \times 10^{-19} \frac{N_{\rm H}}{T_{\rm ex} \Delta v}, \quad (2.26)$$

or

$$N_{\rm H}\,({\rm H\,cm^{-2}}) = 1.82 \times 10^{18} \tau T_{\rm ex} \Delta v = 1.82 \times 10^{18} \tau \int T_{\rm ex} dv\,({\rm K\,km\,s^{-1}}). \quad (2.27)$$

A typical HI cloud has parameters such as the excitation temperature $T_{\rm ex} \sim 100\,{\rm K}$ and $\tau \sim 0.1$, or a peak brightness temperature of about $T_{\rm B} \sim 10\,{\rm K}$, and a velocity dispersion $v_\sigma \sim 10-20\,{\rm km\,s^{-1}}$. These give the column density of a typical HI cloud as

$$N_{\rm H} \sim 2 \times 10^{20}\ ({\rm H\,cm^{-2}}).$$

Given the distance and size of a cloud, the mass can be obtained by

$$M_{\rm HI} = m_{\rm H} \int N_{\rm H} D^2 d\Omega \sim m_{\rm H} N_{\rm H} S, \quad (2.28)$$

where $S \sim \pi r^2 = \pi(\theta D/2)^2$ is the projected area of the cloud with θ being the apparent angular size (diameter). The excitation temperature of HI is typically 100 K for clouds, and 100–1000 K for the diffuse HI gas in the interarm regions and in the halo (thick disk). Typical HI clouds have a size (diameter) of about 10–100 pc, mean density of 10–100 $\rm H\,cm^{-3}$, and a mass of about $10^2 \sim 10^5\,M_\odot$.

The total amount of HI gas in the galaxy can be approximately estimated from the intensity of the 21-cm line emission in the galactic plane. Although individual HI clouds have a velocity dispersion of $\sim$10–20 $\rm km\,s^{-1}$, the line-of-sight integration of many clouds toward a certain direction of the galactic plane yields a much larger velocity width of about $v_\sigma \sim 200\,{\rm km\,s^{-1}}$ because of the galactic rotation of the HI disk. Suppose that a brightness temperature at the galactic plane of about $T_{\rm B} \sim 50\,{\rm K}$ and a line width of $v_\sigma \sim 200\,{\rm km\,s^{-1}}$ have been observed; then the column density along the line of sight is given as

$$N_{\rm H} \sim 10^{22}\ {\rm H\,cm^{-2}}.$$

This yields a mean hydrogen density on the line of sight to be $\sim 0.2\,{\rm H\,cm^{-3}}$. Let us assume that this value is a typical column density when the galaxy is seen edge-on; then we obtain the total mass of the HI disk in the galaxy as

$$M_{\rm HI} \sim m_{\rm H} N_{\rm H} S \sim N_{\rm H}(2R)(2z) \sim 10^9\,M_\odot, \quad (2.29)$$

where $R \sim 15\,{\rm kpc}$ and $2z \sim 400\,{\rm pc}$ are the radius and full-scale thickness of the HI disk.

2.4.2 Distribution of HI Gas

The HI gas clouds are distributed in a disk of thickness $2z \sim 200$ pc, and more diffuse HI gas is distributed in a thicker disk of $2z \sim 400$ pc [2]. The radial distribution is much broader than the H_2 gas: it is distributed from a galactocentric distance of $R \sim 3$ to 30 kpc, peaking at around $R \sim 10$ pc. However, the HI gas is almost lacking in the central 2 kpc region, where the molecular (H_2) gas dominates. Namely, the gas disk of the galaxy comprises two major gas components: the outer HI broad and thick disk with a hole at the center, and the inner thin and dense H_2 disk.

The HI gas most clearly traces spiral arms, and the spiral structure of the galaxy has been mapped by HI observations most intensively. Because the galaxy is seen edge-on, we used the velocity-to-space transformation with which radial velocity profiles are decoded to depths along the lines of sight by using the rotation law (rotation cure) of the disk.

2.5 HI Versus H_2 in the ISM

2.5.1 The HI to H_2 Transition

Concerning the mass, the HI and H_2 gases are the two major components of the ISM, and their mutual fraction is one of the fundamental parameters to describe the ISM condition in the galaxy [4]. We define the molecular fraction here by

$$f_{\rm mol} = \frac{2n_{\rm H_2}}{n_{\rm H} + 2n_{\rm H_2}} = \frac{\rho_{\rm H_2}}{\rho_{\rm H} + \rho_{\rm H_2}}, \tag{2.30}$$

where n_i and ρ_i are the number and mass densities of the species i. In the central few kpc region of the galaxy (near the galactic center) the H_2 gas dominates ($f \sim 1$), whereas in the outer region at $R > 10$ kpc the HI gas dominates ($f_{\rm mol} \sim 0$). A large-scale phase transition from HI to H_2, and vice versa, is clearly seen from their radial distributions.

Locally, the two gas components also avoid each other and are in pressure balance. The lower-density and higher-temperature HI gas is distributed more diffusely than the H_2 gas which is concentrated in higher-density and lower-temperature molecular clouds. The phase transition between HI and H_2 occurs depending on the pressure, metal abundance, starlight field, and the size/mass of a cloud. Here we assume that H_2 molecules are formed from H atoms through catalytic action on surfaces of interstellar grains. The higher are the pressure (number densities of H and grains) and metal abundance (more grains), the higher is the formation rate of H_2 from HI.

On the other hand, H_2 molecules are dissociated by UV radiation from stars, forming HI gas, and therefore, the stronger is the UV radiation (starlight), and the

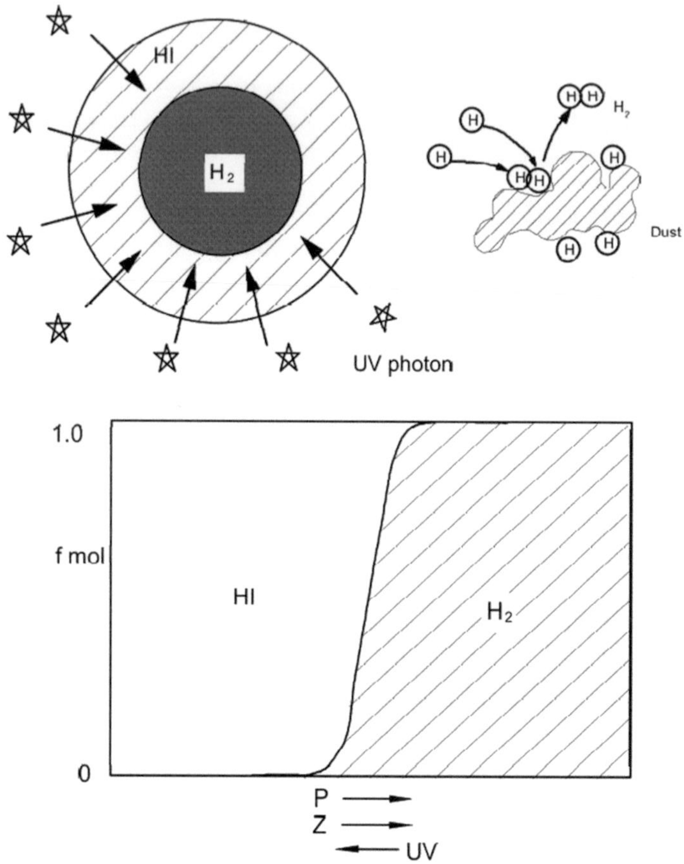

Fig. 2.6 Cloud structure in HI to H_2 phase transition zone, and molecular fraction in HI to H_2 phase transition zone

lower is the formation rate of H_2. Finally, the larger the mass (radius) of the cloud is, the more the obscuration of dissociative UV photons is, leading to a higher fraction of the molecular gas.

In Fig. 2.6 we illustrate the cloud structure and formation and dissociation of molecular gas. Table 2.4 summarizes the dependence of the molecular fraction on the interstellar conditions.

2.5.2 *Molecular Fraction*

Given the mass M of a spherical cloud, which comprises two zones (= central molecular core + HI envelope), we can then calculate the molecular fraction

Table 2.4 HI-to-H_2 transition in the ISM

ISM condition	Increase/decrease	f_{mol} increase/decrease
Pressure (density), P	⇑	⇑
Metal abundance, Z	⇑	⇑
UV photon density, U	⇑	⇓
MC mass (size), $M(r)$	⇑	⇑

$f_{mol}(M : P, Z, U)$ in the cloud for an ISM condition with P, Z, U. By assuming that molecular masses are distributed with a power-law distribution function as

$$f(M)dM \propto M^{-1.5}dM \tag{2.31}$$

with the upper and lower cutoff masses M_{max} and M_{min}, we can further calculate the mass fraction f_{mol} for a given volume of the interstellar space by

$$f_{mol}(P, Z, U) = \int f_{mol}(M : P, Z, U)f(M)dM. \tag{2.32}$$

In the solar neighborhood, the molecular fraction is known to be $f_{mol} \sim 0.2$. The molecular fraction varies with the interstellar condition and, therefore, with the location in the galaxy. The radial variation of f_{mol} implies probing the evolution of interstellar gas in the galaxy scale, which is deeply coupled with the evolution of star formation in the disk: the star formation rate is known to be a function of the molecular gas density. Massive stars among the newly born are the intense source of UV photons. The mass loss and SN explosions from the newborn stars contribute to supply heavy elements to increase Z. Thus the evolution of the ISM and its chemical structure are determined through this cyclic interaction among the HI⇔ H_2 transition, star formation, and heavy element pollution.

2.6 Galactic Scale HI and CO

The major constituents of the ISM in spiral galaxies are the HI and H_2 gases, which make up about a few percent of the total mass [3, 5, 12]. Distributions of HI gas in galaxies have been obtained for many galaxies by mapping with high-resolution observations using large-aperture single dishes such as the 100-m telescope and by interferometers such as the WSRT and VLA. It is, however, only recently that a number of galaxies have been mapped in the molecular lines at mm-wavelengths, particularly in the ^{12}CO($J = 1 - 0$) line emission. The CO line data, such as those obtained at high angular resolutions with the Nobeyama 45-m and IRAM 30-m telescopes, have been used to obtain distributions of the molecular hydrogen gas on the assumption that the CO intensity is proportional to the molecular hydrogen column mass.

It is well known that the HI gas is distributed in the outer region of galaxies, although it is deficient in the central several kpc. On the other hand, the molecular gas is known to be more concentrated in the central several kpc region, often showing a nuclear concentration in the central 1 kpc making a nuclear disk and/or a compact ring. This general tendency of the ISM distribution is the manifestation of a galactic-scale phase transition between HI and H_2 during the evolution of galaxies.

2.6.1 Galactic HI Gas Distributions

The column density of H atoms is obtained by integrating brightness temperature T_{mb} in velocity or in frequency:

$$N_{HI}\ (\mathrm{H\,cm^{-2}}) = 1.83 \times 10^{18} \int_{V_{min}}^{V_{max}} T_B dv \ \ (\mathrm{K\,km\,s^{-1}})$$

$$= 3.88 \times 10^{20} \int_{\nu_{min}}^{\nu_{max}} T_B d\nu \ \ (\mathrm{K\,MHz}). \tag{2.33}$$

The mass is estimated by

$$M_{HI} = m_H \int N_{HI} dxdy = m_H \int N_{HI} D^2 d\Omega, \tag{2.34}$$

where x, y are extents in the galaxy, D is the distance, and Ω is the angular solid angle.

Figure 2.7 shows observed HI distributions in the face of galaxy M51, together with their molecular gas distribution [9]. The neutral hydrogen (HI) gas is widely distributed in the outer region of a galaxy, whereas HI is missing in the central 1 kpc region. The HI distribution in the disk is very broad, and radically extends up to $R \sim 20$–30 kpc.

2.6.2 Galactic CO (H_2) Gas Distribution

The column density of H_2 molecules is obtained by integrating the brightness temperature of the CO line:

$$N_{H_2}\ (\mathrm{H_2\,cm^{-2}}) = C \int_{V_{min}}^{V_{max}} T_B dv \ \ (\mathrm{K\,km\,s^{-1}}), \tag{2.35}$$

where C is the conversion factor for which we often adopt the empirical value obtained for molecular clouds in the galaxy on the assumption of Virial equilibrium of individual clouds. Conventionally, a value $C = 2.8 \times 10^{20}$ is used, and sometimes

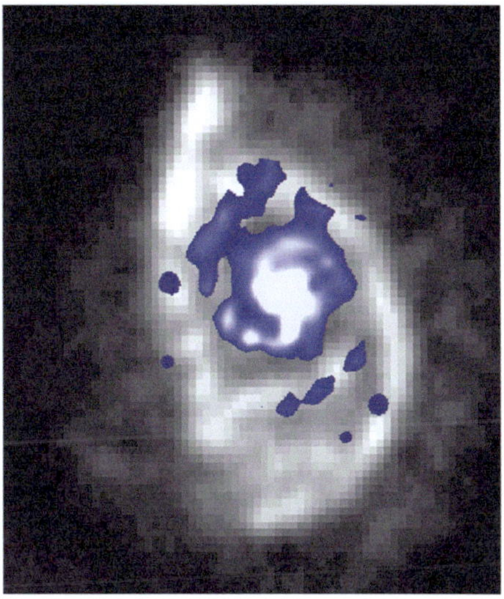

Fig. 2.7 CO intensity distribution overlaid on HI in the face-on galaxy M51 [8, 9]

3.6×10^{20} is adopted. However, the conversion factor has been shown to be clearly dependent on the metallicity of galaxies, and a simple relation between C and metal abundance has been derived.

In contrast to the HI distribution, CO emission comes from more inner and nuclear regions. Figure 2.7 shows the HI and CO intensity distributions for the face-on galaxy M51 [8]. HI is widely spread over the whole galaxy and has a hole (depression) near the nucleus. On the other hand, CO emission has a peak near the nucleus, and the spiral arms are more clearly traced by the CO emission. It is remarkable that HI makes up the outer disk and the outskirts, whereas CO makes up the nuclear disk and inner disk, where star formation is taking place most intensively. The inner molecular disk and outer HI disk appear to be replacing at a certain narrow range of radius at $R \sim 5$–10 kpc, which we later call the molecular front.

2.6.3 *Central Concentration of CO*

Inasmuch as the molecular gas is more concentrated in the inner disk and in the nuclear disk, the emission can be used to obtain the kinematics of the central region, where HI emission is almost missing. Figure 2.8 shows the PV diagram in the CO emission for the edge-on galaxy NGC 891 plotted together with an HI PV diagram. It is impressive that the CO emission has a large-velocity component near the center indicating the existence of a rapidly rotating dense nuclear disk, and HI is not detected in the inner 2 kpc. Rotation curves of the nuclear disk are thus more reliably derived from CO observations, which usually show a steep rise and high velocity inside the galactic bulges.

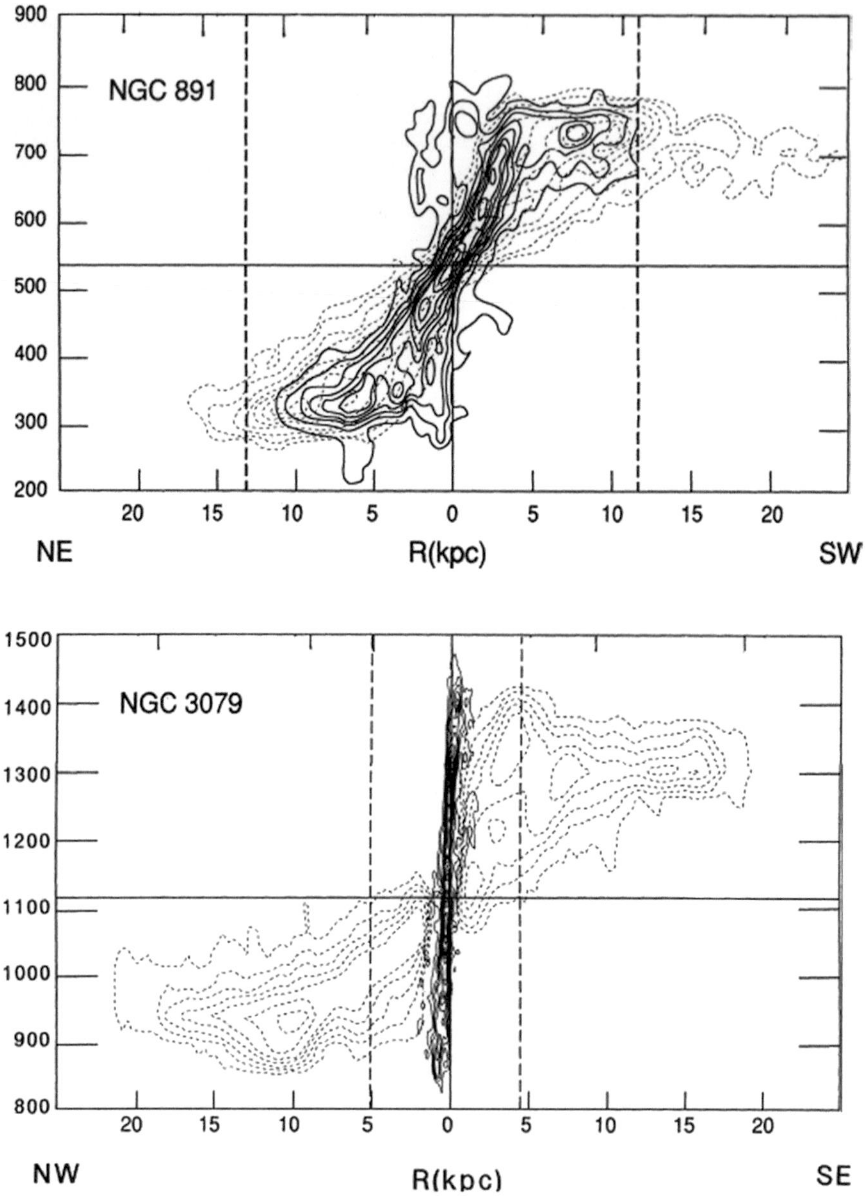

Fig. 2.8 Composite position-velocity (PV) diagrams of the CO (*full contours*) and HI (*dashed contours*) line emissions in the edge-on galaxies, NGC 891 and NGC 3079. *Horizontal lines* indicate the systemic velocities of the galaxies

2.6.4 CO Versus HI in the Position-Velocity (PV) Diagram

The CO emission generally originates in high-density interstellar regions, and the HI in more diffuse interstellar gas, and therefore, a CO-plus-HI PV diagram represents the “true” distribution of interstellar gas. Using the observed PV diagrams, we derive the radial distribution of gas densities of H_2 and HI, and furthermore, the molecular gas fraction.

In Fig. 2.8 we present composite PV diagrams, in which CO and HI PV diagrams are superposed, for the NGC 891, NGC 4565, and NGC 5907. Primarily, these diagrams are useful to obtain rotation properties of the galaxies both in the HI and CO. The CO as well as HI rotation curves already attain the maximum at the galactocentric distance $R \sim 3$–10 kpc, and are followed by a flat rotation of the outer gas. The fact that both the HI- and CO-emitting regions attain the same maximum rotation velocity at $R \sim 5$–10 kpc have been used to argue for a coincidence of the total line profiles in CO and HI and have been used to establish the CO-line Tully-Fisher relation. The diagrams can also be used for deriving dynamical as well as structural information such as regarding a bar potential and a bar-shocked gas ring in the central region. In this chapter we use these diagrams for deriving the molecular and HI gas fractions in the galaxy disks as functions of the distance from the galactic center.

2.7 Radial Distributions of HI and H_2 Densities

The gaseous distributions in the observed galaxies are characterized by a major molecular ring of a few kpc radius and an extended HI outskirts with spiral arms as well as the nuclear disk of molecular gas. Using the observed PV diagrams, we are able to derive radial density distributions both for the HI and molecular hydrogen by applying a deconvolving method as described below.

The HI and H_2 densities in the galaxy disk at a radius R can be obtained using observed intensities and line-of-sight depths by

$$n(\mathrm{HI}) = C_1 b_1^{-1} I_{\mathrm{HI}}/L, \tag{2.36}$$

$$n(\mathrm{H_2}) = C_2 b_2^{-1} I_{\mathrm{CO}}/L, \tag{2.37}$$

where

$$L = 2|R|\sqrt{V_{\mathrm{rot}}^2/V_{\mathrm{min}}^2 - 1}. \tag{2.38}$$

Here, C_1 and C_2 are the conversion factors from the intensities to column densities of H atoms and H_2 molecules, respectively, and are given as $C_1 = 1.82 \times$

10^{18} H cm^{-2}/(K km s^{-1}) and $C_2 = 3.6 \times 10^{20}$ H$_2$ cm^{-2}/(K km s^{-1}). The coefficients b_1 and b_2 are correction factors for the beam dilution in the z direction,

$$b_1 = 2z_{\rm HI}/\theta_1 D \tag{2.39}$$

and

$$b_2 = 2z_{\rm H_2}/\theta_2 D, \tag{2.40}$$

where $2z_{\rm HI} = 250$ pc is the scale thickness of the HI disk, $2z_{\rm H_2} = 150$ pc is the scale thickness of the molecular gas disk, θ_1 and θ_2 are the beam widths of the HI and CO observations, respectively, and D is the distance of the galaxy.

The integrated CO intensity is given by

$$I_{\rm HI} = \int_{V_{\rm min}}^{V_{\rm max}} T_{\rm mb;\,HI} dV \;\; ({\rm K\,km\,s^{-1}}), \tag{2.41}$$

and

$$I_{\rm CO} = \int_{V_{\rm min}}^{V_{\rm max}} T_{\rm mb;\,CO} dV \;\; ({\rm K\,km\,s^{-1}}), \tag{2.42}$$

where $T_{\rm mb;\,HI,\,CO}$ are the main beam temperatures of the HI and CO emissions. The values of $V_{\rm max}$ are taken to be larger than the rotation velocity $V_{\rm rot}$ by 20∼30 km s^{-1} in order to cover all the emissions apparently beyond the terminal velocity due to the velocity dispersion in the interstellar gas.

Figures 2.9 and 2.10 show the distributions of HI and H_2 gas densities plotted against the radius for the Milky Way Galaxy and for the typical Sb spiral galaxies NGC 891 and NGC 4565. The unit of abscissa is H atoms cm^{-3}, where $n({\rm H}) = 2n({\rm H_2})$ for the molecular gas.

The spiral galaxy NGC 891 contains a large amount of molecular gas, clearly showing the R = 4-kpc molecular ring and the nuclear disk. In larger scale the density of the H_2 gas decreases almost exponentially with the radius. On the other hand, the HI density is nearly constant throughout the major part of the disk, as is widely known for normal galaxies. It is also remarkable that the HI gas exhibits a significant depression in the central region at $R < 2$ kpc. We note that the total gas density at $R \sim 8$ kpc is about 0.5 H atoms cm^{-3}, and this is comparable to the value of 0.7 in the solar neighborhood in our galaxy.

NGC 4565 also shows a 5-kpc molecular ring and the nuclear molecular disk is evident, as well as the approximately exponential decrease of the density with the radius. On the other hand, the HI distribution is almost flat in the major disk extending to a large radius such as 20 kpc or more. It is interesting to note that the 5-kpc molecular ring is also associated with a significant enhancement of HI gas.

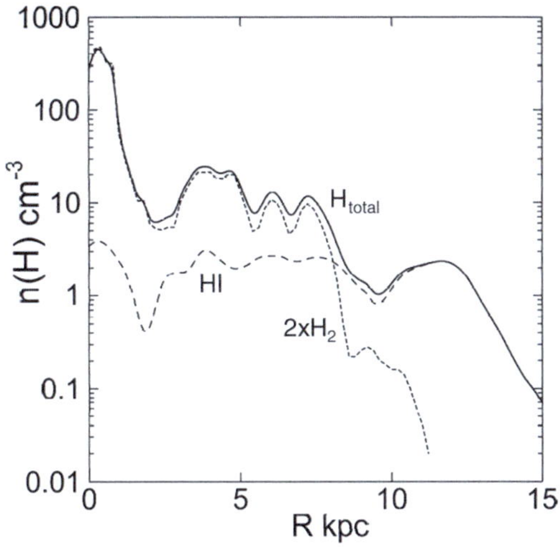

Fig. 2.9 Radial density distributions of HI and H_2 in the Milky Way Galaxy. Note that the ordinate is in logarithmic scaling

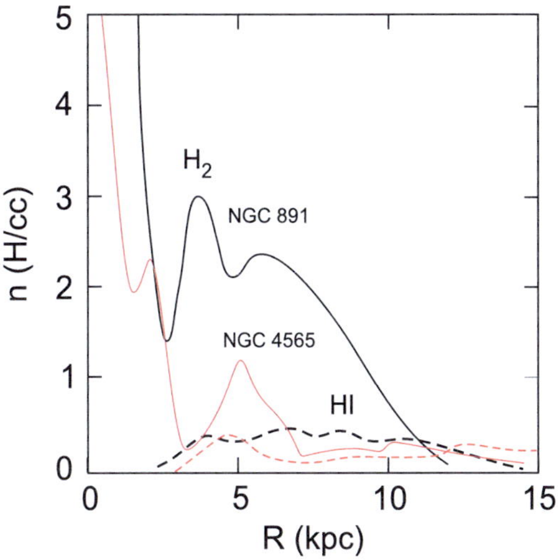

Fig. 2.10 Radial density distributions of HI and H_2 in edge-on galaxies. Note that the ordinate scaling is linear

NGC 5907 shows a significant difference in the H_2 distribution from the other galaxies in the sense that the H_2 gas does not exist in the central 2 kpc region. As we have already seen in the PV diagram, this is because of the lack of the high-velocity component in the central region and the assumption of constant rotation velocity. We cannot ascertain the validity of this assumption for this galaxy, but it is worthwhile to note that there are some galaxies including M31 which show a central depression of interstellar gas.

2.8 Phase Transition Between HI and H_2: Molecular Fraction and the Molecular Front

Using the distributions of the molecular and HI gases, we can derive the radial variation of the molecular gas fraction against the total gas density. We define the molecular fraction by

$$f_{\rm mol} = \frac{\rho_{\rm H_2}}{\rho_{\rm HI} + \rho_{\rm H_2}} = \frac{2n_{\rm H_2}}{2n_{\rm H_2} + n_{\rm H}}. \tag{2.43}$$

Figure 2.11 plots the obtained variation of the molecular fraction as a function of the radius for several galaxies.

2.8.1 *Molecular Front*

All the plots in Fig. 2.11 show that the H_2 gas is dominant in the inner region of the galactic disk, whereas HI is dominant in the outer region. The molecular fraction is almost unity in the central few kpc region, and it decreases drastically at a critical radius. It is remarkable that the sudden decrease in the molecular fraction occurs in positional coincidence with the molecular ring of a few kpc radius. The sudden decrease occurs at $R \sim$ 3–4 kpc in NGC 891, NGC 4565 and NGC 5907, and at $R \simeq$ 4–5 kpc in our galaxy. The molecular fraction, then, decreases to almost zero at $R \sim$ 10 kpc. The galaxy disk is, therefore, separated into two parts at this critical radius: the inner molecular region, where the interstellar gas is almost totally H_2, and the outer HI region, where the gas is almost totally HI.

We stress that the transition from HI to H_2, and vice versa, is taking place within the very narrow radius range, at the inner edge of which the molecular ring is located. We call such a clear border between the two regions of the disk the "molecular front." This front may be the manifestation of a galactic-scale phase transition between HI and H_2, and can be deeply coupled with the evolution of the interstellar gas. It is remarkable that *all* the galaxies studied here have shown similar variations of the molecular fraction.

2.8.2 *Phase Transition Between HI and H_2*

Suppose that a single spherical cloud is composed of two zones (an inner H_2 core and an outer HI envelope) and is embedded in a uniform ISM in a pressure balance. Then, given the mass of the cloud, the transition from HI to H_2, and vice versa, simply depends on the gas pressure (P), the strength of the dissociative-radiation (UV photon) field (U), and the heavy element abundance (Z) related to

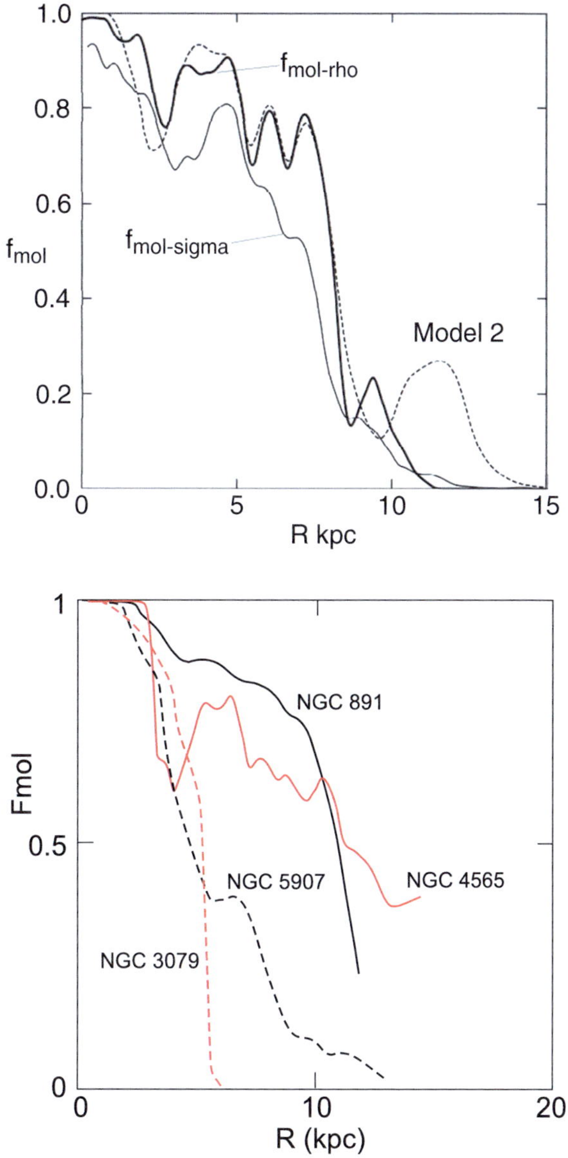

Fig. 2.11 [*Top*] Radial distribution of the surface and volume molecular fraction ($f_{mol-sigma}$, $f_{mol-rho}$) in the galaxy. *Thick line* shows the observation; *thin lines* are model fit based on the observed density distribution. [*Bottom*] Same, but observed surface molecular fraction ($f_{mol-sigma}$) in spiral galaxies. *Dashed lines* are fitted models

the amount of interstellar dust that shields the UV photon. The molecular fraction of the cloud increases if the pressure and/or the heavy-element abundance increase, and if the dissociative UV intensity decreases. The transition is particularly sensitive to the heavy-element abundance and gas pressure. The fraction is also an increasing function of the increasing mass of the cloud. Once a distribution function of cloud masses is given, we can integrate molecular fractions for individual clouds to yield an average value of the molecular fraction (f_{mol}) in a certain area of the gas disk of a galaxy.

It is known that the temperature and velocity dispersion of the molecular as well as HI gas clouds do not vary as steeply with the radius in a galaxy disk. We may, therefore, assume that the pressure is proportional to the total (HI + H_2) gas density, which can be approximately expressed by an exponential disk of scale radius of a few kpc. The radiation field can be assumed to be proportional to the star-formation rate which is a simple power-law function of the gas density. The metal abundance is known to be an exponential decreasing function of the radius. The molecular front phenomenon can be reproduced well by a simple simulation based on such simple assumptions as above in the scope of the phase-transition theory.

References

1. Bolatto, A.D., Wolfire, M., Leroy, A.K.: The CO-to-H_2 conversion factor. ARAA **51**, 207 (2013). [CO to H_2 conversion factor]
2. Dickey, J.M., Lockman, F.J.: HI in the galaxy. ARAA **28**, 215 (1990). [HI gas in the Galaxy]
3. Dickman, R.L., Snell, J.S., Young, J.S. (eds.): Molecular Clouds in the Milky Way and External Galaxies. Springer, Berlin/New York (1988). [HI gas in the Galaxy]
4. Elmegreen, B.G.: The H to H_2 transition in galaxies – totally molecular galaxies. ApJ **411**, 170 (1993). [HI-H_2 transfer]
5. Haynes, M.P., Giovanelli, R., Chincarini, G.L.: The influence of environment on the HI content of galaxies. ARAA **22**, 445 (1984). [HI in galaxies]
6. Kaplan, S.A., Pikel'ner, S.B.: Large-scale dynamics of the interstellar medium. ARAA **12**, 113 (1974). [ISM; review]
7. Maloney, P., Black, J.H.: I(CO)/N(H_2) conversions and molecular gas abundances in spiral and irregular galaxies. ApJ **325**, 389 (1988). [CO-H_2 Converson factor]
8. Nakai, N., Kuno, N., Handa, T., Sofue, Y.: . Distribution and dynamics of molecular gas in the galaxy M51. 1: data and spiral structure. PASJ **46**, 527 (1994). [CO in M51]
9. Rots, A.H., Bosma, A., van der Hulst, J.M., Athanassoula, E., Crane, P.C.: High-resolution HI observations of the Whirlpool Galaxy M51. AJ **100**, 387 (1990). [HI in M51]
10. Shull, J.M., Beckwith, S.: Interstellar molecular hydrogen. ARAA **20**, 163 (1982). [Molecular gas, review]
11. Spitzer, L.: Physical Processes in the Interstellar Medium. Wiley, San Fransisco (1978). [ISM; text book]
12. Young, J.S., Scoville, N.Z.: Molecular gas in galaxies. ARA&A **29**, 581 (1991) [CO in galaxies, review]

Chapter 3
Star Formation and Death

3.1 Mechanism of Star Formation

3.1.1 Sites of Star Formation

Stars are born from interstellar gas by gravitational contraction of molecular cloud cores. Therefore, star-forming sites are where molecular clouds (dark clouds) are crowded. Spiral arms are the major sites of star formation, where the galactic shock wave of gas induced by the density wave compresses the interstellar gas. Particularly, the 4-kpc ring of molecular gas, which is the densest gaseous region of the galaxy, is the most intense star-forming region in the galaxy. The central region of the galaxy, where the nuclear molecular disk is found, is also a place of intense star formation.

Figure 3.1 shows the distribution of HII regions in the galaxy overlaid on the distribution maps of interstellar HI gas (shown in red) and molecular gas (green). HII regions are distributed approximately in spatial coincident with the molecular gas disk. Also they trace the spiral arms. The number density of HII regions in the galactic plane represents the star-formation rate (SFR) as discussed below.

Among newly born stars, massive stars (OB stars) evolve rapidly, and explode as type II supernovae (SN) when they die. (Type I SN are for stars less massive, but greater than the solar mass.) Through the SN explosion, the stars supply kinetic energy to the ISM as well as pollute it with heavy elements. The lifetime of OB stars are $10^6 \sim 10^7$ yr, so that the type II SN explosion occurs relatively close to the places where they were born.

The timescale of star formation is approximately given by the contraction time (free-fall time $\sim 10^6$ yr) of the cloud core, which is short enough compared to the galactic rotation period (10^8 yr). The lifetime of OB stars is also short, and the luminosity of OB stars greatly exceeds the B and V band luminosity of the

---

The original version of this chapter was published without printing chapter abstract. The abstracts for each chapter are updated in the front matter in the current version of the book. An erratum to this chapter can be found at https://doi.org/10.1007/978-981-10-3445-9_7.

Y. Sofue, *Galactic Radio Astronomy*, Lecture Notes in Physics 935,
DOI 10.1007/978-981-10-3445-9_3

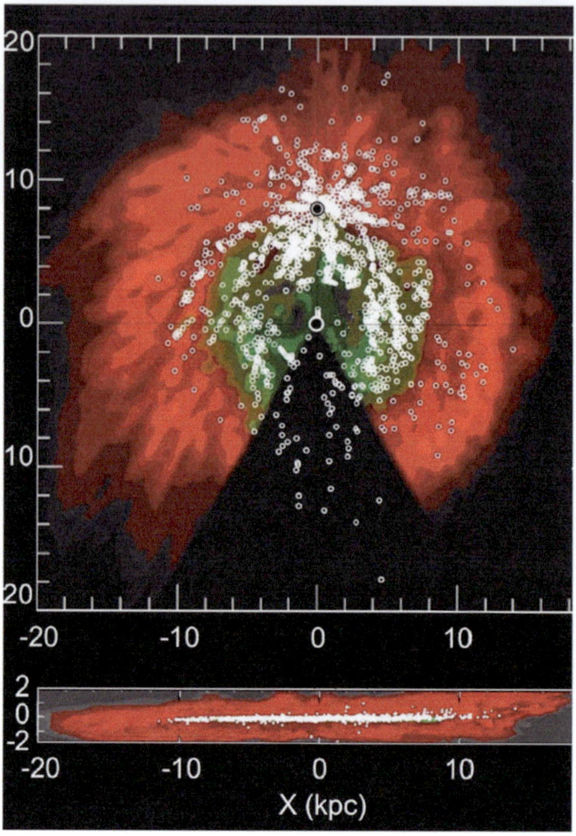

Fig. 3.1 Distribution of HII regions [6], which are star-forming regions in the galaxy, overlaid on the gas density distribution map, where *green* is molecular gas and *red* is HI gas [10, 11]. Stars are born from the molecular gas

other newly born stars. Because of these facts, the spiral arms shine most intensely compared to the interarm region. This gives rise to the high contrast between the optical spiral arms and other regions on *B* and *V* band photographs of galaxies, impressing one with the spiral structure.

3.1.2 Schmidt's Law

Stars are formed from interstellar gas globally obeying the scaling relation called Schmidt's law [7]. The star-formation rate (*SFR*) is defined by

$$SFR = \frac{d\rho_{\rm stars}}{dt}, \tag{3.1}$$

and the star-formation efficiency (*SFE*) by

$$SFE = \frac{d\rho_{\rm stars}}{\rho_{\rm gas} dt}, \tag{3.2}$$

where $\rho_{\rm stars}$ and $\rho_{\rm gas}$ are the densities of stars and gas in the interstellar space of a certain volume, say within 100 pc to 1 kpc square region.

The SFR can be determined from observations by counting OB associations and HII regions by dividing them by their assumed lifetime:

$$SFR({\rm OB}) \sim \frac{n_{\rm OB}}{\tau_{\rm OB}}, \tag{3.3}$$

with $\tau_{\rm OB} \sim 10^7$ yr, or

$$SFR({\rm HII}) \sim \frac{n_{\rm HII}}{\tau_{\rm HII}}, \tag{3.4}$$

with $\tau_{\rm HII} \sim 10^6$ yr. The SFR can also be obtained from the H_α intensity, and from the UV photon density that can be estimated from thermal radio emissivity.

The SFR is higher where gas density is higher, and vice versa. Hence, the SFR can be reasonably related to the gas density by Schmidt's law:

$$SFR \propto \rho_{\rm gas}^{\alpha}, \tag{3.5}$$

or

$$SFE \propto \rho_{\rm gas}^{\alpha-1}. \tag{3.6}$$

If the star formation is stochastic, the power-law index would be around $\alpha \sim 1$. On the other hand, if it is related to a collision process of gas clouds, the power would be $\alpha \sim 2$.

The star-formation law can be obtained by counting the number of HII regions and OB associations, and so on, as Eqs. (3.3) and (3.4) averaged in a certain volume of the interstellar space. However, the gas density can be measured from HI line emission or from CO line observations. The power index α is then estimated by plotting $\rho_{\rm OB}(\rho_{\rm HII})$ against the gas density $\rho_{\rm gas}$.

Figure 3.2 shows the Schmidt law in the galaxy, or a logarithmic plot of the SFR against gas density, exhibiting a linear correlation on the log-log plane, and therefore, SFR can be expressed by a power law of gas density. The law shown here is expressed by $SFR \propto \rho_{\rm H_2}^{0.67\pm0.02}$ for molecular gas, $\propto \rho_{\rm HI}^{2.71\pm0.07}$ for HI gas, and $\propto \rho_{\rm total}^{2.20\pm0.08}$ for the total gas.

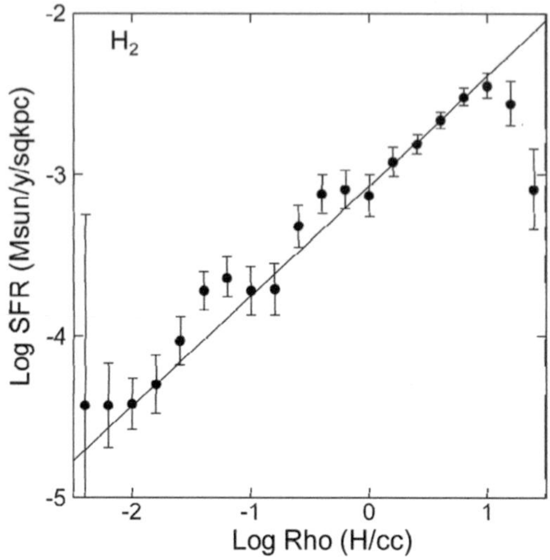

Fig. 3.2 Schmidt's law in the Milky Way. A *plot* of the star-formation rate *SFR* ($\propto$ number density of star-forming regions/lifetime) against the molecular (H_2) gas density in the galactic disk

3.1.3 Birth of Stars

Stars are born from interstellar gas by gravitational contraction [15]. The most efficient site of formation is molecular clouds, particularly cores in giant molecular clouds (GMC). Fragmentation of molecular clouds and GMCs into smaller and denser clouds is triggered by various instabilities in the ISM [1]. The major mechanism is the gravitational instability, whereas other types of instabilities work simultaneously and are supplementary.

The fragments further contract gravitationally to form "protostellar" clouds and then evolve to "protostars." The timescale of the contraction is on the order of gravitational free-fall time given by $t \sim 1/\sqrt{4\pi G\rho}$, which is often called the Jeans time. The timescale is on the order of $\sim 10^6$ yr in GMC cores.

The critical mass for growth of the instability is determined by the critical wavelength or size of the contracting cloud, and is called the Jeans mass and Jeans wavelength, respectively. In molecular clouds the Jeans mass is on the order of $\sim 0.1\,M_\odot$ for a gas of density 10^4 $H_2\,cm^{-2}$ and temperature 10 K. The Jeans mass gives a lower limit of the mass of protostellar clouds.

Note that the Jeans mass is the lower limit of the gravitationally unstable clouds, and therefore, clouds of mass larger than this critical mass can form stars. This means that stars born from a molecular cloud have a distribution in a wide range of masses greater than the Jeans mass.

An ensemble of stars born from a single cloud or a group of clouds form a cluster of stars. Young groups of newly born stars are often observed as OB associations and open clusters of stars, which are the typical population I stellar objects in the galactic disk.

Globular clusters, comprising 10^5 stars, are a quite different type of clusters, born in the far past from much larger clouds during the formation of the galaxy. They are one of the oldest objects of population II.

3.1.4 *Initial Mass Function (IMF)*

The distribution of mass or number density of stars as a function of the mass of stars is called the initial mass function (IMF) or simply the mass function [14]. The IMF is estimated by counting stars in a young cluster of stars or by counting local stars of various types (various masses).

The IMF $f(M)$ $(M_\odot M_\odot^{-1})$ is defined by the mass (or mass density) of the cluster or ensemble of stars per unit stellar mass range (per unit volume)

$$f(M) = M\frac{dN(M)}{dM} = \frac{dN(M)}{d\ln M}(M_\odot M_\odot^{-1}) \propto M^\mu, \tag{3.7}$$

where $N(M)$ (count) is the accumulated number (or number density) of stars with stellar mass less than or equal to M. An alternative function is the number (or number density) of stars per unit stellar mass range defined by

$$g(M) = \frac{dN(M)}{dM} \propto M^{\mu-1}(\text{count } M_\odot^{-1}). \tag{3.8}$$

The index defined as the slope of the logarithmic plot of $f(M)$ against M,

$$\mu = d\log f/d\log M, \tag{3.9}$$

is one of the most fundamental quantities in star formation and evolution, and has been observed to be $\mu \simeq -1.35 \sim -1.5$. The IMF having index $\mu \sim -1.35$ is often called Salpeter's mass function.

There are lower and upper masses of stars in a cluster of stars, which are usually $M_{\rm min} \sim 0.1\,M_\odot$ and $M_{\rm max} \sim 50\,M_\odot$, respectively. The total mass of the cluster is calculated by

$$M_{\rm total} = \int_{M_{\rm min}}^{M_{\rm max}} f(M)dM \propto \frac{1}{\mu}(M_{\rm max}^{\mu+1} - M_{\rm min}^{\mu+1}) \tag{3.10}$$

Because μ and $1+\mu$ are both negative and $M_{\rm min} \ll M_{\rm max}$, this formula reduces to

$$M_{\rm total} \sim M_{\rm min}N(M_{\rm min}). \tag{3.11}$$

The mass of a cluster is, therefore, dominated by low mass stars. This applies to larger-scale stellar systems from globular cluster to galaxies. Namely, the stellar

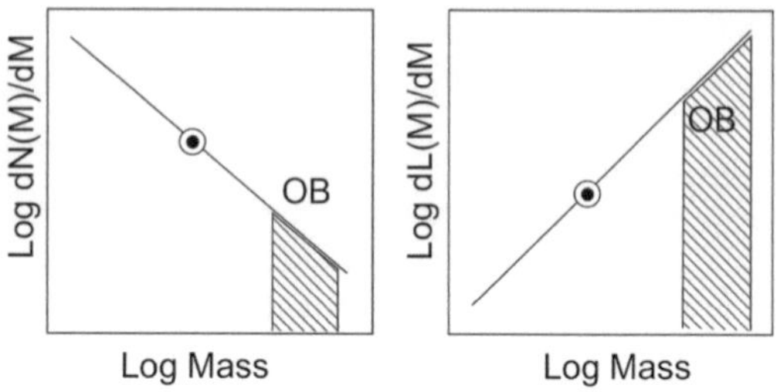

Fig. 3.3 Mass function (mass of stars per unit stellar-mass range per unit volume) and luminosity function (luminosity of stars per unit stellar-luminosity range per unit volume)

mass of a galaxy is dominated by old, population II stars. On the other hand, larger-mass stars contribute more to the luminosity than lower-mass stars. Hence, the luminosity is mainly determined by the high-mass stars. Figure 3.3 schematically shows the mass function and luminosity function of a cluster or galaxy.

3.1.5 The Virial Theorem

The gravitational equilibrium, expansion, or contraction of interstellar clouds and fluctuations of the medium can be expressed in terms of the Virial theorem. The moment of inertia of a system with density ρ is given by

$$I = \int \rho\, \boldsymbol{r}^2 dxdydz. \tag{3.12}$$

The system is in equilibrium when

$$\frac{d^2 I}{dt^2} = 2K + \Phi_{\mathrm{G}} + E_{\mathrm{M}} + 3(\gamma - 1)U_{\mathrm{T}} = 0, \tag{3.13}$$

where

$$K = \frac{1}{2}\int \rho v^2 dxdydz \tag{3.14}$$

is the kinetic energy;

$$\Phi_{\mathrm{G}} = -\frac{1}{2}\int \frac{\rho\rho'}{|\boldsymbol{r} - \boldsymbol{r}'|} dx'dy'dz' \tag{3.15}$$

is the potential energy;

$$E_{\mathrm{M}} = \frac{1}{8\pi}\int B^2 dxdydz \tag{3.16}$$

is the magnetic energy density; and

$$U_{\mathrm{T}} = \frac{1}{\gamma - 1}\int p\, dxdydz \tag{3.17}$$

is the internal (thermal) energy with p as the pressure. For example, if a cloud is in equilibrium with its kinetic energy because of its internal velocity dispersion (turbulence) balancing the gravitational energy, we have

$$\frac{d^2I}{dt^2} = 2K + \Phi_{\mathrm{G}} = 0. \tag{3.18}$$

If

$$\frac{d^2I}{dt^2} > 0 \tag{3.19}$$

the system expands, whereas if

$$\frac{d^2I}{dt^2} < 0 \tag{3.20}$$

it gravitationally collapses.

3.1.6 Gravitational Contraction of Clouds

Suppose a gas cloud of density ρ, radius r, and an internal velocity dispersion v_σ which is approximately equal to the sound velocity $v_\sigma \sim c_{\mathrm{s}}$ (Fig. 3.4). The mass is given by $M = \frac{4\pi}{3} r^3 \rho$, and the gravitational energy

$$\Phi = \frac{GM^2}{r}. \tag{3.21}$$

The kinetic energy is given by

$$K = \frac{1}{2}Mc_{\mathrm{s}}^2. \tag{3.22}$$

If the cloud is in a Virial equilibrium, it is stable:

$$\frac{GM^2}{r} \sim Mc_{\mathrm{s}}^2, \tag{3.23}$$

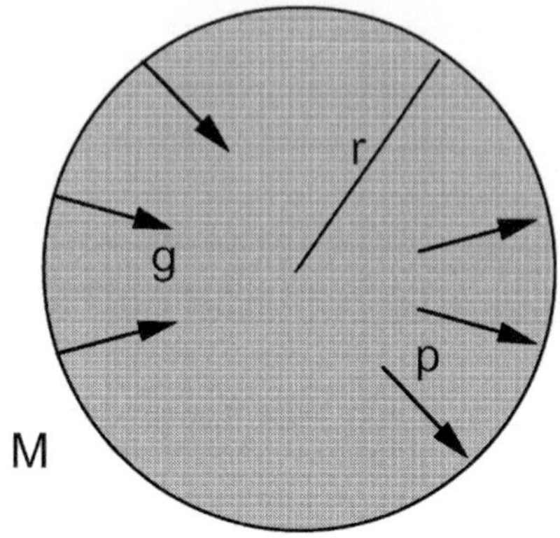

Fig. 3.4 Gravitational contraction of a spherical gas cloud

or

$$r \sim \frac{c_s}{\sqrt{4\pi G\rho}}. \tag{3.24}$$

(See the next subsection for a more accurate treatment of instability for the factor 4π.) However, the cloud is gravitationally unstable and collapses if the gravity dominates:

$$\frac{GM^2}{r} > Mc_s^2, \tag{3.25}$$

or if

$$r > \frac{c_s}{\sqrt{4\pi G\rho}}. \tag{3.26}$$

The timescale of gravitational contraction is given by the free-fall time:

$$\frac{d^2r}{dt^2} \sim \frac{r}{t^2} \sim \frac{GM}{r^2}, \tag{3.27}$$

or

$$t \sim \sqrt{\frac{r^3}{GM}} \sim \frac{\sqrt{3}}{\sqrt{4\pi G\rho}} = \frac{4.6 \times 10^7}{\sqrt{n\ (\mathrm{H\ cm^{-3}})}}\ (\mathrm{yr}). \tag{3.28}$$

This is also given by the timescale that is necessary for information to transfer and cross the cloud at the sound velocity:

$$t \sim \frac{r}{c_s} \sim \frac{1}{\sqrt{4\pi G\rho}}. \tag{3.29}$$

The scale length

$$\lambda_J \sim c_s t \sim \frac{c_s}{\sqrt{4\pi G\rho}} \tag{3.30}$$

is called the Jeans wavelength. Because the sound velocity is related to the temperature $c_s = \sqrt{\gamma RT}$ (R is the gas constant), the Jeans wavelength can also be expressed as

$$\lambda_J \sim \sqrt{\frac{\gamma RT}{4\pi G\rho}} \sim \sqrt{\frac{\gamma R}{4\pi G m_H}}\sqrt{\frac{T}{n}} \sim 7\sqrt{\frac{T\ (\mathrm{K})}{n\ (\mathrm{H\,cm^{-3}})}}\ (\mathrm{pc}) \tag{3.31}$$

The mass given by

$$M_J \sim \rho\lambda_J^3 \sim 23(T/1\mathrm{K})^{3/2}(n/1\mathrm{H\,cm^{-3}})^{-1/2}\,M_\odot \tag{3.32}$$

is called the Jeans mass. The Jeans wavelength, mass, and free-fall time for typical interstellar clouds as well as for gaseous constituents in the universe are given in Table 3.1. The masses estimated for molecular gas and cloud cores are typical for stars.

Table 3.1 Jeans wavelength, mass, and time (free-fall time)

Cloud	ρ (m_H cm^{-3})	T (K)	λ_J (pc)	M_J ($M_\odot$)	t_J (y)	
GMC core	$\sim 10^5$	$10-30$	~ 0.1	$0.1-1$	3¥ $\times 10^5$	To stars
GMC	$\sim 10^3$	$10-100$	~ 1	~ 10	3¥ $\times 10^6$	To stars
HI cloud	~ 100	10^2-10^3	~ 10	$\sim 10^3$	–	–
Intercloud gas	~ 1	10^4	–	–	–	–
Intracluster	$\sim 10^{-3}$	10^{6-7}	–	–	–	–
Universe	$\sim 10^{-6}$	10^6	–	–	–	–

3.1.7 Jeans Instability

Hydrodynamical equations describing a gas of density ρ are given as

$$\rho\frac{d\boldsymbol{v}}{dt} = -\nabla p + \rho\nabla\Phi, \tag{3.33}$$

$$\frac{d\rho}{dt} + \rho\nabla\cdot\boldsymbol{v} = 0, \tag{3.34}$$

$$\nabla^2\Phi + 4\pi G\rho = 0. \tag{3.35}$$

Here,

$$\frac{d}{dt} = \frac{\partial}{dt} + \boldsymbol{v}\cdot\nabla = 0. \tag{3.36}$$

Here Eqs. (3.33), (3.34), and (3.35) are the equations of motion, fluid continuity, and Poisson's equation, respectively.

Consider a case when the density, pressure, and velocity are expressed as a superposition of a uniform component and a small perturbation (Fig. 3.5):

$$\rho = \rho_0 + \delta\rho, \tag{3.37}$$

$$p = p_0 + \delta p, \tag{3.38}$$

$$\boldsymbol{v} = \boldsymbol{v}_0 + \delta\boldsymbol{v}. \tag{3.39}$$

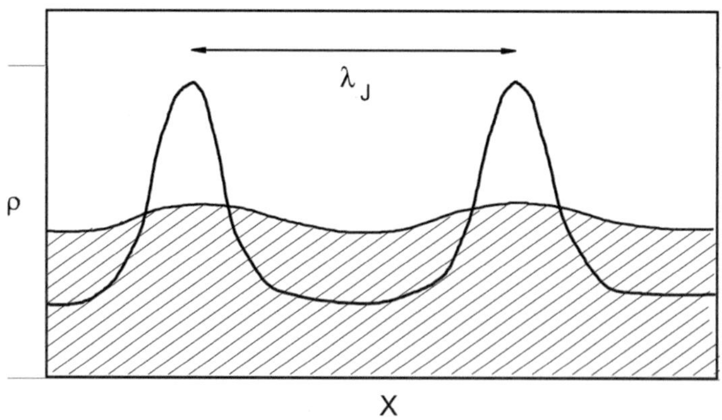

Fig. 3.5 Gravitational instability (Jeans instability) of a density perturbation of a uniform gas

Remembering that ρ_0, p_0, and $\boldsymbol{v}_0$ satisfy Eqs. (3.33), (3.34), and (3.35), we can rewrite the above equations as

$$\rho_0 \frac{d\delta \boldsymbol{v}}{dt} = -\nabla \delta p + \rho_0 \nabla \delta \Phi, \tag{3.40}$$

$$\frac{\partial \delta p}{\partial t} = -\rho_0 \nabla \cdot \delta \boldsymbol{v}, \tag{3.41}$$

and

$$\nabla^2 \delta \Phi = -4\pi G \delta \rho. \tag{3.42}$$

Knowing that the pressure is related to the density by the adiabatic form, as $p \propto \rho^\gamma$, and that the sound velocity is related to ρ and p as

$$c_{\rm s}^2 = \frac{dp}{d\rho} \simeq \frac{\delta p}{\delta \rho} = \gamma \frac{p}{\rho}, \tag{3.43}$$

we have

$$\frac{\delta p}{p} = \gamma \frac{\delta \rho}{\rho}, \tag{3.44}$$

or

$$\delta p = c_{\rm s}^2 \delta \rho. \tag{3.45}$$

On the other hand, Eqs. (3.40) and (3.41) reduce to

$$\rho \frac{\partial}{\partial t} \nabla \cdot \delta \boldsymbol{v} = -c_{\rm s}^2 \nabla^2 \delta \rho + \rho \nabla^2 \delta \Phi. \tag{3.46}$$

Eliminating the terms including Φ from Eqs. (3.42) and (3.46), we obtain

$$\frac{\partial^2 \delta \rho}{\partial t^2} = c_{\rm s}^2 \nabla^2 \delta \rho + 4\pi G \rho_0 \delta \rho. \tag{3.47}$$

This is a wave equation, and can be solved by introducing the exponential function

$$\delta \rho = \delta \rho_0 e^{i(\boldsymbol{k} \cdot \boldsymbol{r} + \omega t)}, \tag{3.48}$$

to yield the dispersion relation

$$\omega^2 = c_s^2 k^2 - 4\pi G\rho_0 = c_s^2(k^2 - k_J^2), \tag{3.49}$$

with

$$k_J = \frac{1}{c_s}\sqrt{4\pi G\rho_0} \tag{3.50}$$

called the Jeans wave number, or

$$\lambda_J = \frac{1}{k_J} = \frac{c_s}{\sqrt{4\pi G\rho_0}} \tag{3.51}$$

called the Jeans wavelength.

When $k < k_J$ or $\lambda > \lambda_J$, we have $\omega^2 < 0$, and the dispersion relation gives a growing wave: the instability grows, and the system is said to be unstable. The growth time is given by

$$t \sim \frac{1}{|\omega|} = \frac{1}{\sqrt{4\pi G\rho_G - c_s^2 k^2}}. \tag{3.52}$$

For a small wave number $k \ll k_J$, or for a large wavelength $\lambda \gg \lambda_J$, we have

$$t \sim \frac{1}{\sqrt{4\pi G\rho_0}}, \tag{3.53}$$

which is approximately the same as the free-fall time.

On the other hand, if the wave number is large enough, and the wavelength is small ($k > k_J$ or $\lambda < \lambda_J$, $\omega^2 > 0$), the dispersion relation leads to an oscillation solution. This solution corresponds to a propagating compression wave: the small perturbation cannot grow, but is smeared out with a sound wave. The instability, then, does not grow, and the system is stable.

3.1.8 Thermal Instability

Even if the gravity of a cloud is too small to compress the gas, the cloud can be compressed by the surrounding gas pressure [4]. In a steady state, the cooling and heating are in balance, thus the cloud shape is kept unchanged. If the cloud is given a density perturbation, the cooling rate increases and it may happen that the cooling overcomes the heating. This causes a decrease in the internal temperature and pressure, which is followed by compression of the cloud by the surrounding pressure. The compression further accelerates cooling, and, therefore, accelerates pressure suppression. This process is called thermal instability. It occurs when the cooling time t_{cool} is shorter than the traveling time of the sound wave t_{sound}:

$$t_{\mathrm{cool}} < \tau_{\mathrm{sound}}, \tag{3.54}$$

where

$$t_{\mathrm{cool}} = \mathscr{R}T/\Lambda, \tag{3.55}$$

and

$$\tau_{\mathrm{sound}} = \lambda/c_{\mathrm{s}}. \tag{3.56}$$

Here, Λ is the cooling rate of the gas, λ is the size of the cloud, and $c_{\mathrm{s}} = (\gamma p/\rho)^{1/2}$ is the sound speed. Figure 3.6 illustrates the growth of density of a cloud by the cloud's thermal instability, and Fig. 3.7 shows the cooling functions.

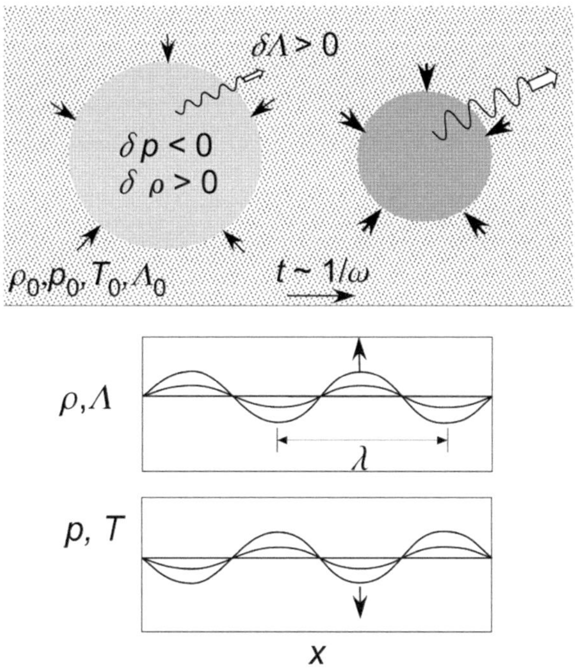

Fig. 3.6 Thermal instability causing compression of a cloud

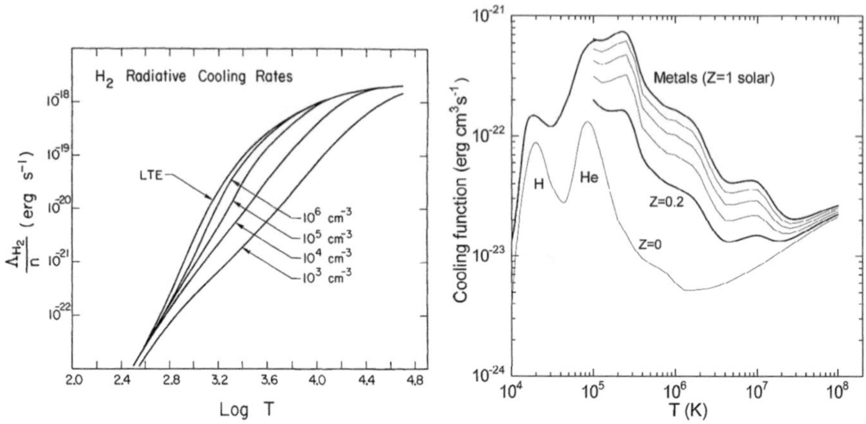

Fig. 3.7 (*Left*) Cooling function representing Λ/n of a low-temperature gas by molecular line radiation [16]. (*Right*) Cooling function representing Λ/n^2 of hot plasma for various metallicity

Hydrodynamical equations describing the gas of density ρ are given as

$$\rho\frac{d\boldsymbol{v}}{dt} + \nabla p = 0, \tag{3.57}$$

$$\frac{d\rho}{dt} + \rho\nabla\boldsymbol{v} = 0, \tag{3.58}$$

$$\frac{1}{\gamma-1}\frac{dp}{dt} - \frac{\gamma}{\gamma-1}\frac{p}{\rho}\frac{d\rho}{dt} + \rho(\Lambda - \mathscr{H}) - \nabla(K\nabla T) = 0, \tag{3.59}$$

and

$$p = \frac{\mathscr{R}}{\mu}\rho T. \tag{3.60}$$

Here, the first two equations are the continuity of mass and momentum, but the gravity term is neglected. Equation (3.59) represents the variation of the internal (thermal) energy density caused by compression, cooling, heating, and thermal conductivity.

As in the case for the Jeans instability, we express the variables as a superposition of those in equilibrium and small perturbation with the exponential function form as

$$x = x_0 + \delta x_1\, e^{\omega t + i\,\boldsymbol{k}\cdot\boldsymbol{r}}. \tag{3.61}$$

Note that the frequency ω is included here as a growth term, instead of $i\omega$ as in the Jeans instability. Therefore if ω is real, the instability grows: either it is compressed or expands. If ω is imaginary, the cloud's density oscillates.

Inserting these expressions about $\delta\rho$, δT, δp, and $\delta \boldsymbol{v}$ into the four basic equations and eliminating the variables, we obtain an equation that relates ω and k as follows.

$$\omega^3 + \omega^2 c_s (k_T + k^2/k_K) + \omega c_s^2 k^2 + \frac{c_s^2 k^2}{\gamma}(k_T - k_\rho + k^2/k_K) = 0. \tag{3.62}$$

Here,

$$k_\rho = \frac{\mu(\gamma - 1)}{\mathscr{R} c_s} \frac{\rho_0}{T_0} \frac{\partial \Lambda}{\partial \rho}, \tag{3.63}$$

$$k_T = \frac{\mu(\gamma - 1)}{\mathscr{R} c_s} \frac{\partial \Lambda}{\partial T}, \tag{3.64}$$

and

$$k_K = \frac{\mathscr{R} c_s \rho_0}{\mu(\gamma - 1) K}. \tag{3.65}$$

If we can neglect the conductivity and $\mu = 1$, we obtain

$$\omega^3 + \omega^2 \frac{\gamma - 1}{t_{\rm cool}} \frac{\partial \ln \Lambda}{\partial \ln T} + \omega c_s^2 k^2 + \frac{(\gamma - 1) c_s^2 k^2}{\gamma t_{\rm cool}} \left(\frac{\partial \ln \Lambda}{\partial \ln T} - \frac{\partial \ln \Lambda}{\partial \ln \rho} \right) = 0. \tag{3.66}$$

This dispersion relation, or the characteristic equation, can be used to find the domain of instability to satisfy $\omega > 0$ in the (k, ω) space.

3.1.9 Rayleigh-Taylor Instability

In astrophysics various types of instability are known to occur in the ISM in addition to the gravitational thermal instabilities. Typical instabilities are the Rayleigh-Taylor (RT) instability, Kelvin-Helmholtz (KH) instability, Parker instability, and magnetorotational instability (MRI).

Rayleigh-Taylor instability is a well-known phenomenon occurring on the contact surface between oil and water under gravity, when water with higher density is put above the oil with lower density. The contact surface begins to become wavy, and, as the waves grow, the water clumps drops into the oil.

The same phenomenon takes place in the interstellar medium when a molecular cloud is accelerated by high-pressure (high-temperature) gas such as caused by

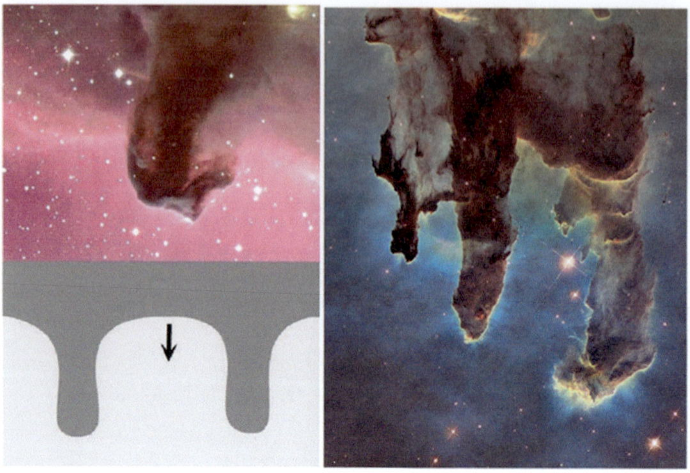

Fig. 3.8 Rayleigh-Taylor instability in dark nebulae accelerated by expanding HII gas and radiation pressure by newly born stars, feeling gravity (acceleration toward the *bottom* in the figure). The illustration shows RT instability at the water-oil contact surface in gravity. The color images show the Horsehead Nebula in Orion (APOD 2000, http://apod.nasa.gov/apod/ap090721.html, Credit & Copyright: Nigel Sharp (NOAO), KPNO, AURA, NSF (National Science Foundation)), and the Eagle Nebula (AAPOD 2015, https://apod.nasa.gov/apod/ap160424.html, Credit & Copyright: Laszlo Francsics [Orion Nebula])

an expanding HII region. The pushed molecular gas feels acceleration toward the low-density region, as if there were gravity in the same direction. Thus, the higher-density fluid is put over the lower-density fluid against the gravity (acceleration), causing Rayleigh-Taylor instability.

Figure 3.8 illustrates Rayleigh-Taylor instability occurring along a contact surface between a molecular gas and HII region. The photo shows the famous Horsehead Nebula in Orion, which is the place of growing RT instability. Here, the high-density molecular gas (dark cloud) at the top is pushed from below by the expanding HII region with lower density driven by the star-forming region located farther in the bottom.

3.1.10 Kelvin-Helmholtz Instability

Kelvin-Helmholtz instability is the driving mechanism of waves on the water surface of the ocean by the wind of the air. It occurs at the contact surface of two fluids (here, water and air) in shear motion.

In the interstellar medium, it occurs when a cloud suffers from a different flow (wind) of the surrounding gas. Such flow occurs often along the surface of a growing pillar of dark cloud formed by the RT instability, where the HII gas is flowing along the pillar, or the pillar is growing/moving into the HII region. Figure 3.9 shows

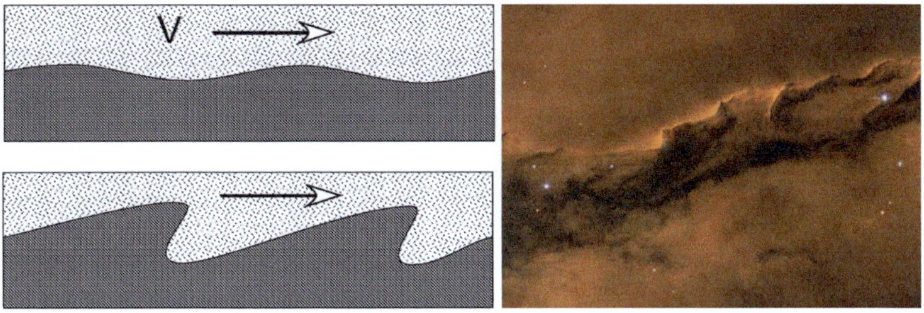

Fig. 3.9 Kelvin-Helmholtz instability and excited waves on the surface of a pillar of dark cloud in the Eagle Nebula induced by an HII gas wind blowing from the *left* to *right* (APOD 2005, http://apod.nasa.gov/apod/ap050425.html, Credit: The Hubble Heritage Team, STScI/AURA, ESA, NASA)

an image of the Eagle Nebula in part, showing the wavy surface driven by KH instability due to a flow from left to right in the figure.

In larger scales, KH instability is triggered by the shearing motion of disk gases at neighboring flow tubes in differential galactic rotation. The induced wavy structures are observed as fins (or spurs) growing radially from spiral arms, as shown in for the case of M51.

3.1.11 Parker Instability (Magnetic Inflation)

The interstellar space is penetrated by a galactic magnetic field, to which cosmic rays are confined. The interstellar gas includes more or less thermal free electrons that are trapped in the magnetic field, and hence even neutral interstellar gas and clouds are also trapped in the field lines. Namely, the interstellar gas and magnetic fields are "frozen in" to each other. The entire magnetized interstellar medium filled by gas and cosmic rays is confined to the galactic disk by the galactic gravitational force toward the disk plane. However, the gas and cosmic rays cannot move perpendicularly to the field lines, but can slip parallel to the field lines.

Suppose that field lines run parallel to the galactic plane and the gas and cosmic rays are confined to the field as illustrated in Fig. 3.10. If a sinusoidal perturbation takes place in the shape of the field lines, the gas at the top of the concave field lines is pulled to the galactic plane by gravity, and cosmic rays are kept to be distributed uniformly along the line. The cosmic-ray pressure at the top is kept the same, but the gas density and the gravity by the gas to push the field lines down decrease. This produces buoyancy near the top, and the field lines are inflated toward the halo. Accordingly, the gas slips down along the field lines toward the disk plane to push the field lines near the valley toward the plane. These actions accelerate the growth of the buoyancy, and the field lines are inflated to form wavy and, further, loop-shaped arches in the halo.

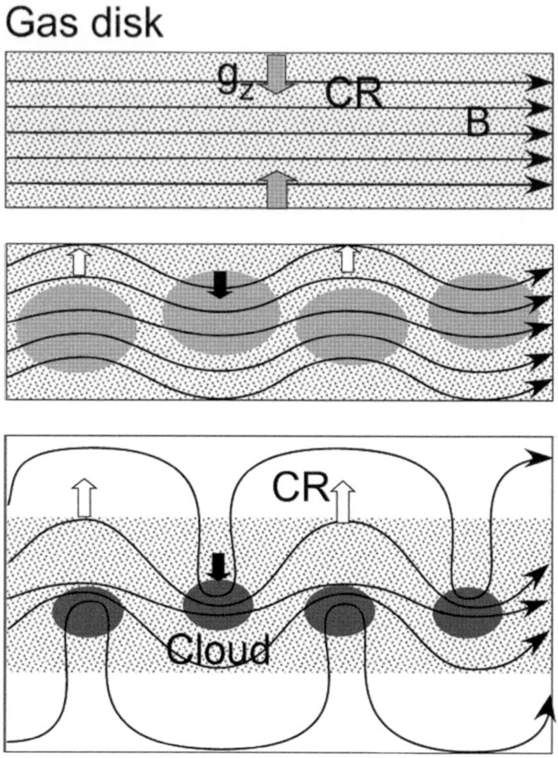

Fig. 3.10 Parker instability in a gas disk filled with cosmic rays and magnetic field with gravity perpendicular to the disk

This mechanism is called Parker instability, which was originally proposed as a mechanism to excite the dynamo action in the solar atmosphere, and to explain the magnetic loops in the solar corona.

Such inflated magnetic waves, arches, and loops filled with interstellar gas are often recognized in the ISM distribution maps of the Milky Way. Figure 3.10 shows an example of a nearby Parker arch of HI and molecular gas penetrated by the magnetic field and a schematic illustration of two-dimensional magnetic lines of force after numerical MHD simulation of the nonlinear growth of the instability.

3.2 Environment of Star Formation

3.2.1 Triggering of Cloud Compression

A galactic shock wave is induced by the density wave of a stellar disk, and is the largest scale mechanism to compress the ISM in the galaxy. The gas is compressed at the shock to form gaseous spiral arms, where the HI gas is transferred to molecular

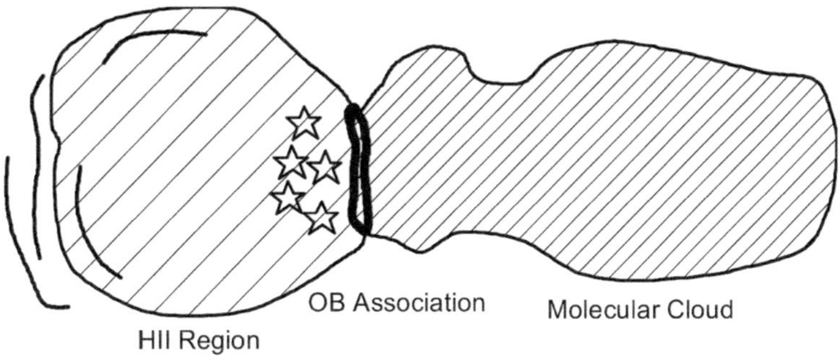

Fig. 3.11 Star formation in molecular clouds

clouds. In the central region of the galaxy, an oval potential of the inner stellar disk induces a stronger shock of gas, and the gas is rapidly accreted to the nuclear disk, where molecular clouds are formed and intense star formation occurs.

Stars are born through fragmentation of dense cores of molecular clouds and GMCs because of the gravitational instability of density perturbations (fluctuation) [4]. Such density fluctuations are triggered by local compression of gas either due to a shock wave penetration from outside or stochastic collisions among turbulent eddies inside the cloud. In either case, collisions between gas clouds and cloudlets (fragments) are the major sources of interstellar gas compression. Typical examples of collision phenomena are observed during mutual collisions of turbulent cloud eddies inside a cloud, collisions of two MCs or GMCs, an encounter with an expanding shell of supernova remnants, an encounter by expanding HII regions (Fig. 3.11), injection of jets from the central region, and so on.

3.2.2 Shock Wave

The shock wave is described by two discontinuous compressive gaseous flows 1 and 2 [9] (Fig. 3.12). Let the temperature, velocity, density, pressure, and the internal energy of the two gases be $T_1,\ u_1,\ \rho_1,\ p_1, U_1 = \frac{1}{\gamma-1}p_1$ and $T_2,\ u_2,\ \rho_2,\ p_2, U_2 = \frac{2}{\gamma-1}p_2$, respectively. The basic equations describing the gases are: The equation of continuity:

$$\rho_1 u_1 = \rho_2 u_2, \tag{3.67}$$

The continuity of the momentum:

$$p_1 + \rho_1 u_1 u_1 = p_2 + \rho_2 u_2 u_2, \tag{3.68}$$

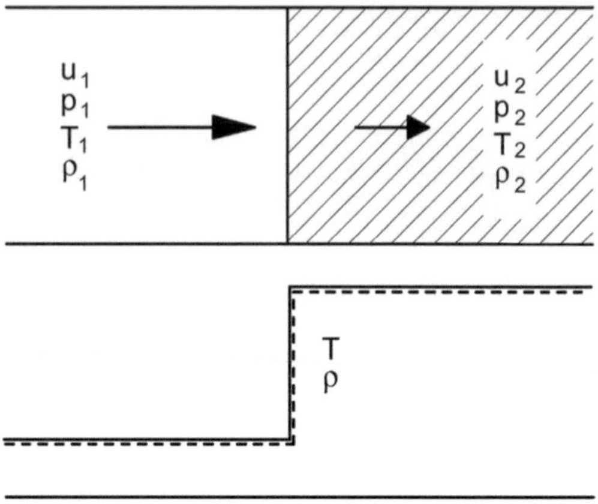

Fig. 3.12 A shock wave caused by supersonic collision of two gases

and the continuity of the energy:

$$u_1 p_1 + u_1 \left(\frac{1}{2} \rho_1 u_1^2 + U_1 \right) = u_2 p_2 + u_2 \left(\frac{1}{2} \rho_2 u_2^2 + U_2 \right), \tag{3.69}$$

or

$$u_1^2 + \frac{2\gamma}{\gamma - 1} \frac{p_1}{\rho_1} = u_2^2 + \frac{2\gamma}{\gamma - 1} \frac{p_2}{\rho_2}. \tag{3.70}$$

Equation (3.67), (3.68), and (3.69) are solved to give

$$\frac{p_2}{p_1} = \frac{2\gamma}{\gamma - 1} M^2 - \frac{\gamma - 1}{\gamma + 1}, \tag{3.71}$$

and

$$\frac{u_2}{u_1} = \frac{\rho_1}{\rho_2} = \frac{\gamma - 1}{\gamma + 1} + \frac{2}{\gamma + 1} \frac{1}{M^2}. \tag{3.72}$$

Here,

$$M = \frac{u_1}{c_1} = \sqrt{\frac{\rho_1 u_1^2}{\gamma p_1}} \tag{3.73}$$

is the Mach number, and

$$c_1 = \sqrt{\gamma R T_1} = \sqrt{dp/d\rho} \tag{3.74}$$

is the sound velocity. When

$$u_1 \gg c_1 \ (M \gg 1) \tag{3.75}$$

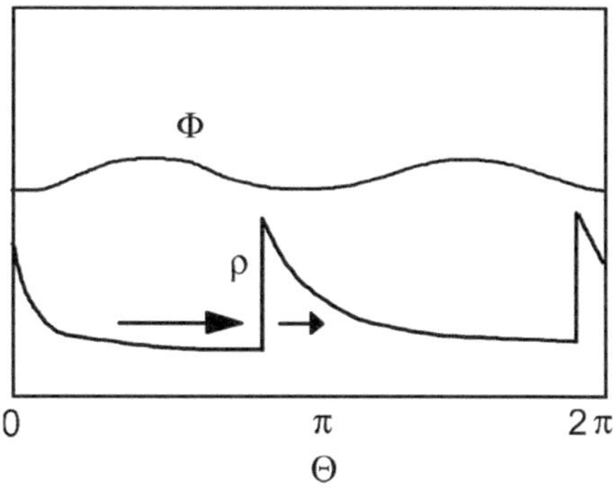

Fig. 3.13 Compression of gas in the arm by galactic shock wave

the gas is supersonic, and the solution reduces to

$$\frac{\rho_1}{\rho_2} \simeq \frac{\gamma - 1}{\gamma + 1}. \tag{3.76}$$

If the gas is adiabatic with the adiabatic index $\gamma = 5/3$, the supersonic shock compression gives

$$\frac{\rho_2}{\rho_1} \simeq 4. \tag{3.77}$$

If the gas is isothermal, or the temperature does not depend on the density and the pressure is proportional to the density ($p \propto \rho$; $\gamma = 1$), we have

$$\frac{\rho_2}{\rho_1} = M^2. \tag{3.78}$$

The interstellar gas during galactic shock compression (Fig. 3.13) or during a cloud-cloud collision is almost isothermal, because the cooling time is shorter compared to the compression timescale. In a galactic shock wave, the velocity of inflowing HI gas into a density-wave potential is on the order of $u_1 \sim 20$–$30\,\mathrm{km\,s^{-1}}$ and the sound velocity is about $c_1 \sim 5$–$10\,\mathrm{km\,s^{-1}}$, which yields a Mach number $M \sim 2$–6 and the compression of $\frac{\rho_2}{\rho_2} = M^2 \sim 4$–$36$. If two molecular clouds of internal sound velocity ~ 0.1–$1\,\mathrm{km\,s^{-1}}$ collide at a velocity $u_1 \sim 5\,\mathrm{km\,s^{-1}}$, the compression is as high as $\rho_2/\rho_1 \sim 25$ to 10^3.

According to Schmidt's law for the star-formation rate and the gas density, the SFR can be expressed as $SFR \propto \rho_{\mathrm{gas}}^{\alpha}$ with $\alpha \sim 1$–2. The SFR, therefore, depends on the Mach number as

$$SFR \propto \frac{\rho_2}{\rho_1}^{\alpha} \sim M^{2\alpha}. \tag{3.79}$$

This relation, particularly when $\alpha \sim 2$, indicates that the star formation is most intense in the galactic shock region, and thus is in the region near the galactic center, or in colliding gas clouds at a supersonic velocity.

3.2.3 Formation of Molecular Clouds

The formation of molecular clouds, particularly of giant molecular clouds, occurs during the large-scale compression of interstellar gas in the spiral arms. The mean density of ISM in the disk of the galaxy is $n \sim 1\,\mathrm{H\,cm^{-3}}$. The gas is compressed in the galactic shock by a compression rate of 10–30, or $n \sim 10$–$100\,\mathrm{H\,cm^{-3}}$. The Jeans wavelength in a compressed gaseous arm with $n \sim 10\,\mathrm{H\,cm^{-3}}$ and sound velocity $c_s \sim 10\,\mathrm{km\,s^{-1}}$ is then given by

$$\lambda_J \sim \frac{c_s}{\sqrt{4\pi G\rho_{gas}}} \sim 100\ \mathrm{pc}. \tag{3.80}$$

The Jeans mass is, therefore,

$$M_J \sim \rho\lambda_J^3 \sim 1.5\times 10^5\,M_\odot. \tag{3.81}$$

The contraction time of this cloud is given by

$$t \sim 1/\sqrt{G\rho} \sim 10^7\ \mathrm{yr}, \tag{3.82}$$

which is comparable to the crossing time of the cloud through an arm.

3.2.4 Why Spiral Arms Are Bright

Molecular clouds are condensed along the galactic shock region in spiral arms (Fig. 3.13). Molecular clouds contain dust, therefore the dense gaseous arm is observed as a dark lane (Fig. 3.14. Stars are born in groups from dense molecular clouds. The mass function of born stars (mass M) (number of stars per unit volume per unit stellar mass) has the form

$$f(M)dM \propto M^\mu dM \tag{3.83}$$

with $\mu \simeq -1 \sim -2$, and the often-quoted value is $\mu \sim -1.5$ [8, 14]. This function is often called the initial mass function, and indicates that less massive stars dominate in number, and fewer massive stars are born compared to smaller stars.

The luminosity of a single star is approximately proportional to the mass as

$$L \propto M^{3.5} \tag{3.84}$$

The luminosity function per unit of the stellar mass is then written as

$$\phi(M)dM = Lf(M)dM \propto M^{3.5}M^\mu dM \sim M^2 dM. \tag{3.85}$$

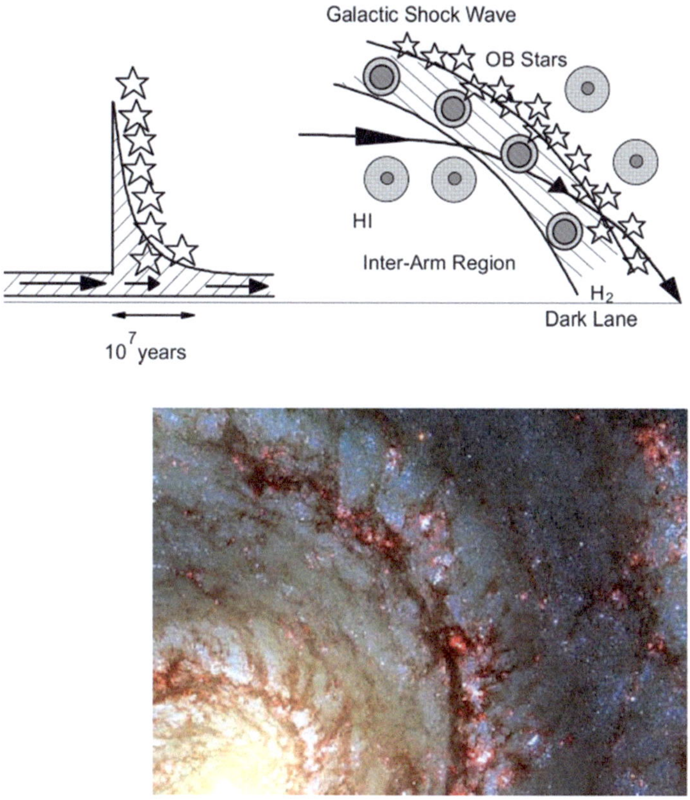

Fig. 3.14 Why spiral arms shine. Gas clouds are compressed along the galactic shock wave, and star formation is enhanced. Young bright stars (OB stars) are thus oriented along the arm, but they evolve rapidly and do not survive to cross the arm. The color image is a part of the bright arm in the spiral galaxy M51 with the rotation direction mirrored (APOD 2008, http://apod.nasa.gov/apod/image/0806/M51HST-GendlerM.jpg)

The luminosity function per unit of stellar luminosity is then related to the star luminosity L as

$$\phi(L)dL = Lf(M)dL \propto L^{1+\mu/3.5}dL \sim L^{0.7}dL. \tag{3.86}$$

This shows that the luminosity per unit volume in the galactic disk is dominated by the light from most massive (luminous) stars, although their number is much less compared to smaller (less massive) stars. The lifetime of a star can be related to the mass and luminosity as

$$t \sim M/L \sim M^{-2.5} \tag{3.87}$$

Massive stars evolve much more rapidly than smaller stars, and the lifetime of O- and B-type stars is as short as $\sim 10^5 - 10^7$ yr. The star-forming region soon after the formation of a star cluster shines predominantly because of OB stars, and the color is "blue," emitting intense UV radiation. As time elapses, massive stars evolve more rapidly than less massive stars do, and the OB stars explode to die and disappear. Thus, the luminosity of the region decreases rapidly within a few 10^7 yr, and the color becomes "redder."

By galactic rotation the interstellar gas flows through a spiral arm, and the gas is compressed by the galactic shock during the passage through the arm. Molecular clouds form there, and are observed as the dark lane. OB stars are born from the clouds and shine in the downstream region of the dark lane for a few 10^7 yr, and are observed as a bright spiral belt. OB stars cannot survive until they leave the arm into the interarm region, which takes about 10^8 yr. For these reasons, the belt of OB stars along the galactic shock wave is particularly bright in the disk. On the other hand, the luminosity per unit volume in the interarm region is far less than that along the bright arm.

3.3 HII Regions

3.3.1 Ionization Sphere: Strm̈gren Sphere

A young stellar association soon after its formation contains massive stars of O and B types. The OB stars emit strong UV radiation, and lionize the ambient interstellar gas. Therefore, a star-forming region is usually associated with an ionized gas, called an HII region (Fig. 3.15).

Let $N_{\mathrm{UV}} \simeq L_{\mathrm{uv}}/h\nu$ be the number (production rate) of photons with wavelength shorter than $\lambda_0 = 912\mathring{A}$ emitted by an OB-star cluster, where L_{uv} is the luminosity of the cluster and $h\nu$ is the typical UV photon energy. The UV photons ionize hydrogen atoms in the ambient gas of density n_{H}. On the other hand, the ionized hydrogen recombines at a rate of $n_e n_i \alpha_r$, and absorbs UV photons at this same rate.

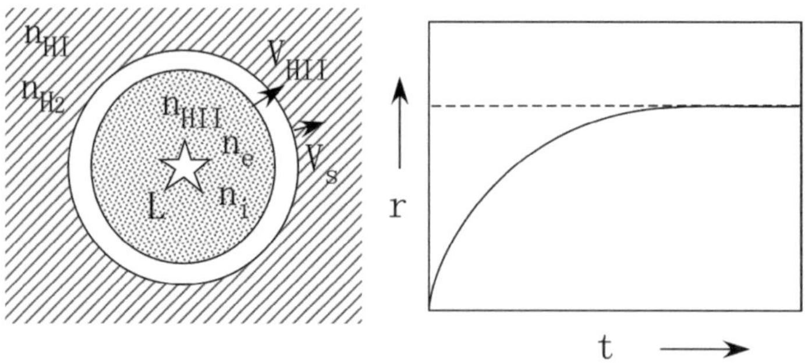

Fig. 3.15 Expansion of ionization sphere of radius r around an OB-star cluster

If the gas is composed of pure hydrogen, the rate of ionization and recombination in a sphere (radius r) of ionized gas around this OB star cluster is written as

$$4\pi r^2 n_{\mathrm{H}} \frac{dr}{dt} = N_{\mathrm{UV}} - \frac{4\pi}{3} n_{\mathrm{e}} n_{\mathrm{i}} \alpha_{\mathrm{r}} r^3. \tag{3.88}$$

Here, the left-hand side is the rate of photoionization by UV radiation, and the right-hand side is the rate of photo production by the OB stars and absorption by recombined atoms within the sphere. The shell of radius r is called the ionization front. The solution of this differential equation is given by

$$\frac{4\pi r^3}{3} = \frac{N_{\mathrm{UV}}}{\alpha_{\mathrm{r}} n_{\mathrm{e}} n_{\mathrm{i}}} \left(1 - e^{-\frac{\alpha_{\mathrm{r}} n_{\mathrm{i}} n_{\mathrm{e}}}{n_{\mathrm{H}}} t}\right). \tag{3.89}$$

In the initial phase of ionization formation, when the radius of the sphere is small enough, the recombination term in the right-hand side of Eq. (3.88) is negligible:

$$\frac{4\pi r^3}{3} = V = \frac{N_{\mathrm{UV}}}{n_{\mathrm{H}}} t, \tag{3.90}$$

or

$$r \simeq \left(\frac{3 N_{\mathrm{UV}} t}{4\pi n_{\mathrm{H}}}\right)^{1/3}. \tag{3.91}$$

In the final phase of the expansion rate of the sphere (dr/dt) becomes zero, and the ionization and recombination balance each other:

$$\frac{4\pi r^3}{3} \simeq \frac{N_{\mathrm{UV}}}{n_{\mathrm{i}} n_{\mathrm{e}} \alpha_{\mathrm{r}}}, \tag{3.92}$$

or

$$r \simeq \left(\frac{3 N_{\mathrm{UV}}}{4\pi n_{\mathrm{i}} n_{\mathrm{e}} \alpha_{\mathrm{r}}}\right)^{1/3} \sim 4.1(\mathrm{pc}) \left(\frac{n_{\mathrm{H}}}{100\,\mathrm{cm}^{-3}}\right)^{-2/3} \left(\frac{\alpha_{\mathrm{r}}}{4 \times 10^{-13}\,\mathrm{cm}^{-3}\mathrm{s}^{-1}}\right)^{-1/3}, \tag{3.93}$$

where the electron n_{e} and ion n_{i} densities are assumed to be equal to the hydrogen atom number density n_{H}.

This ionization sphere around OB stars in which the equilibrium between ionization and recombination is realized is called a "Strömgren's sphere." The recombination rate α_{r} is related to the temperature T and metal abundance Z as

$$\begin{aligned} \alpha_{\mathrm{r}} &= 2A \left(\frac{2kT}{\pi m_{\mathrm{e}}}\right)^{1/2} \frac{h\nu_0}{kT} Z^2 \phi \left(\frac{h\nu_0 Z^2}{kt}\right) \\ &= \left(\frac{2.06^{-11} Z^2}{T^{1/2}}\right) \phi \left(\frac{h\nu_0 Z^2}{kt}\right) \mathrm{cm}^3\mathrm{s}^{-1}. \end{aligned} \tag{3.94}$$

Here,

$$A = \frac{2^4}{3^{3/2}} \frac{he^2}{m_e c^3} = 2.105 \times 10^{-22}\,\mathrm{cm}^2 \tag{3.95}$$

($\propto \sigma$ = cross section of recombination) and ν_0 is the frequency at $\lambda_0 = 912\mathring{A}$, or $h\nu_0/k = 158{,}000\,\mathrm{K}$. The ϕ is a slow function of T and Z (of order unity at $T > 10^5$ K and $\sim 10 - 100$ at $T \sim 10^5 - 10^4$ K).

Suppose a single O5 star, or equivalently a cluster including 5 B0 stars, surrounded by a uniform hydrogen gas of density $n_{\mathrm{H}} \sim n_{\mathrm{i}} \sim 10\,\mathrm{H\,cm^{-3}}$. The luminosity of the star(s) is $\sim 3 \times 10^5 L_\odot = 10^{39}\,\mathrm{erg\,s^{-1}}$, emitting $N_{\mathrm{UV}} \sim 5.2 \times 10^{49}$ UV photons $\mathrm{s^{-1}}$. For a recombination coefficient of $\alpha_{\mathrm{r}} \sim 4^{-13}\,\mathrm{cm^3 s^{-1}}$, we obtain the radius of the Strömgren sphere as $r \sim 20\,\mathrm{pc}$.

3.3.2 Expanding Ionization Front

In the above expression of a spherical HII region we have ignored the flow (or expansion) of the gas inside the sphere. However, the ionized gas has been heated to $\sim 10^4$ K, much higher than the temperature of the surrounding gas (~ 100 K for HI, and $\sim 10 - 100$ K for H_2 gas). This implies that the ionization sphere is not in a pressure equilibrium with the ambient gas: it expands because of the higher internal pressure.

In the initial expansion phase, the expansion velocity of the sphere $v = dr/dt$ and the radius r are related to the internal energy E as

$$E \sim \frac{1}{2} M v^2 \sim \frac{1}{2} \frac{4\pi r^3}{3} m_{\mathrm{H}} n_{\mathrm{H}} v^2, \tag{3.96}$$

where $M_{\mathrm{s}} = \frac{4\pi}{3} r^3 m_{\mathrm{H}} n_{\mathrm{H}}$ is the plowed mass in the sphere from the ambient gas. The internal energy is related to the integrated photon energy that has been spent for ionization of the hydrogen gas minus radiated energy by the recombination:

$$E \sim L_{\mathrm{uv}} t - \int \frac{4\pi r^3}{3} h\nu n_{\mathrm{i}} n_{\mathrm{e}} \alpha_{\mathrm{r}} dt. \tag{3.97}$$

Alternatively, because the temperature of the HII region is known to be maintained at $\sim 10^4$ K due to radiative cooling as well as expansion, this energy can be written simply as

$$E \sim \gamma M_{\mathrm{s}} R T \sim \gamma \frac{4\pi r^3}{3} n_{\mathrm{H}} k T. \tag{3.98}$$

Using Eqs. (3.96) and (3.97), we obtain

$$v \sim \sqrt{2\gamma kT} \sim c_s. \tag{3.99}$$

This implies that the sphere expands at the sound velocity of the internal gas: $v \sim c_s \sim 10\,\mathrm{km\,s^{-1}}$ for $T \sim 10^4$ K. Namely, the expansion velocity is large enough compared to the sound velocity in the ambient gas ($c_{s0} \sim 0.3\,\mathrm{km\,s^{-1}}$ for H_2 gas; $\sim 1\,\mathrm{km\,s^{-1}}$ for H gas). The expansion is supersonic with a Mach number $M \sim 10$ and, therefore, a strong shock wave forms at the interface of the surrounding neutral gas and the expanding HII region (sphere).

In a later phase when a Strömgren sphere is realized, the ionization front at radius r expands faster (at a velocity $V_{\rm ion}$) than the expanding HII sphere of radius with a velocity $v_{\rm HII}$ and $n_e \sim n_i \sim \rho_{\rm HII}$. From the Strömgren sphere condition we have

$$r_{\rm ion}^3 n_e n_i \sim r_{\rm HII}^3 \rho_{\rm HII} \sim \text{const}, \tag{3.100}$$

which leads to

$$V_{\rm ion} = dr_{\rm ion}/dt \sim -\frac{2r_{\rm ion}}{3\rho_{\rm HII}}\frac{d\rho_{\rm HII}}{dt}. \tag{3.101}$$

The HII sphere expands following

$$r_{\rm HII}^3 \rho_{\rm HII} \sim \text{const}, \tag{3.102}$$

to yield

$$v_{\rm HII} = dr_{\rm HII}/dt \sim -\frac{r_{\rm HII}}{3\rho_{\rm HII}}\frac{d\rho_{\rm HII}}{dt}. \tag{3.103}$$

Because $r_{\rm HII} \sim r_{\rm ion}$, from Eqs. (3.101) and (3.103) we have:

$$v_{\rm HII} \sim \frac{1}{2} V_{\rm ion}. \tag{3.104}$$

The expanding sphere of the HII gas forms an expanding shell at velocity $V_s \sim vII \sim 1/2vion$, and the ambient HI gas of density $\rho_{\rm HI}$ (velocity = 0) is accumulated. From the continuity condition of the shock front we have

$$\rho_{\rm HI} V_s^2 = p_{\rm HII} + \rho_{\rm HII}\left(\frac{1}{2}V_s\right)^2. \tag{3.105}$$

Here, $p_{\rm HII}$ is the pressure of the HII region and is related to the density and sound velocity of the HII gas $c_{\rm HII}$ as ($\gamma \sim 1$);

$$p_{\rm HII} \sim \rho_{\rm HII} c_{\rm HII}^2. \tag{3.106}$$

Then we obtain an expression that relates the expanding velocity of the HII shell with the sound velocity and ambient HI density as

$$V_s \sim c_{\mathrm{HII}} \sqrt{\frac{\eta}{1 - \frac{1}{4}\eta}}, \tag{3.107}$$

where η is the density ratio of the HII and HI regions.

$$\eta = \rho_{\mathrm{HII}}/\rho_{\mathrm{HI}}. \tag{3.108}$$

3.3.3 *Shock Compression of Ambient Gas*

The above equations also apply for an HII region expanding into a molecular cloud, in which case the HI density can be read as H_2 density. When the HII region expands into an HI cloud or into a molecular cloud, the HII density is small enough compared to the ambient gas density; $\rho_{\mathrm{HII}} \ll \rho_{\mathrm{HI}}$ or $\rho_{\mathrm{HII}} \ll \rho_{\mathrm{mc}}$, and the expansion velocity is simply expressed as

$$V_s \sim c_{\mathrm{HII}} \sqrt{\rho_{\mathrm{HII}}/\rho_{\mathrm{HI}}} \;\; (c_{\mathrm{HII}} \sqrt{\rho_{\mathrm{HII}}/\rho_{\mathrm{mc}}}). \tag{3.109}$$

In a normal HII region, we may assume $c_{\mathrm{HII}} \sim 10\,\mathrm{km\,s^{-1}}$ (for $T \sim 10^4\,\mathrm{K}$) and $\rho_{\mathrm{HII}} \sim 10^4\,\mathrm{H\,cm^{-3}}$. If the shell expands into an HI cloud of density $\rho_{\mathrm{HI}} \sim 100\,\mathrm{H\,cm^{-3}}$, the expansion velocity is at $V_s \sim 3\,\mathrm{km\,s^{-1}}$, and is supersonic. If the shell expands into a molecular cloud of density $\rho_{\mathrm{mc}} \sim 10^3\,\mathrm{H\,cm^{-3}}$, where the sound velocity is about $0.3\,\mathrm{km\,s^{-1}}$, the expansion is at $1\,\mathrm{km\,s^{-1}}$, and is still supersonic.

From these estimations we learn that an HII region expanding into an HI cloud or into a molecular cloud produces a shock wave and compresses the neutral gas to form a high-density shocked layer. If we assume that the gas is isothermal, the compression of the gas at the shocked layer is on the order of $M^2 \sim 10$ for $c_s \sim 0.3\,\mathrm{km\,s^{-1}}$ ($T \sim 10\,\mathrm{K}$) and $V_s \sim 1\,\mathrm{km\,s^{-1}}$. Such a strong compression of the neutral gas triggers further contraction of protostellar gas clouds.

3.4 Sequential Star Formation

3.4.1 *Propagation of Shock Compression by an HII Region*

Once a cluster of stars including OB stars is formed in or near the surface of a molecular cloud, an HII region develops around the OB stars. The HII region expands, and produces a shock-compressed layer in the molecular cloud with compression by a factor of $\sim$10. The growth time ($\sim\sqrt{1/4\pi\rho}$) of the compressed layer is smaller than that of the ambient gas, and the gravitational instability grows more rapidly, resulting in an enhancement of further fragmentation of the layer into protostellar clouds [9, 18].

Thus the compression by an expanding HII region results in a further formation of stars in the molecular cloud, giving rise to a new HII region caused by newly born OB stars (Fig. 3.16). This new HII region, then, compresses the cloud to form another compression layer. In this way, the "star-forming layer" propagates into the cloud interior. This process is called "sequential star formation" in a molecular cloud [3].

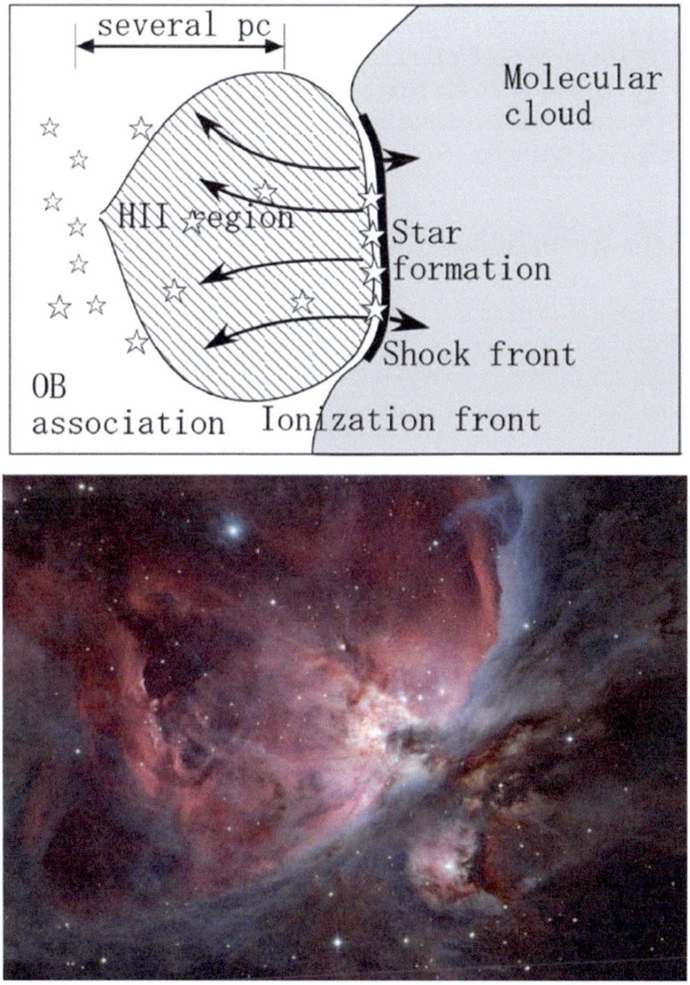

Fig. 3.16 Shock compression of a molecular gas cloud by an expanding HII region, compared with a photo of the Orion Nebula (APOD 2015, http://apod.nasa.gov/apod/ap150107.html, Image Credit: NASA, ESA, and The Hubble Heritage Team, STScI/AURA)

3.4.2 *Various SF Regions*

3.4.2.1 Orion Region

The Orion molecular cloud, particularly its core, is an intermediate class of star-forming region, where infrared sources indicating an on-going star formation have been found. It is known that this star-forming region is associated with an HII region, called Ori A (in radio) and the Orion Great Nebula (in optical), whose age is approximately 10^6–10^7 yr. The Orion star-forming region is further associated with a young stellar association involving stars of age $\sim 10^7$ yr several pc away. Another association of stars with an age $\sim 10^8$ has been found in a farther location at about a few tens of pc. Furthermore, all these objects are surrounded by a diffuse HII shell of radius $\sim$100 pc as seen in the Hα emission. These facts show that the place of star formation has shifted toward the present molecular cloud region in the past 10^8 yr. Namely, a sequential star formation has occurred in the Orion molecular cloud for the past 10^8 yr, and is still on-going.

3.4.2.2 M16–M17 Region

M17 is an intense HII region associated with giant molecular clouds that extend for about 100 pc along the galactic plane. The region is supposed to be the place where the sequential star formation is on-going.

M16 is an older HII region than M17 with an extended Hα nebula and associated thermal radio emission region. A giant shell of diameter about 50 pc is found in the thermal radio emission and is also surrounded by an HI gas shell expanding at a velocity of about 20 km s^{-1}.

3.4.2.3 Rosette Nebula

The Rosette Nebula (Fig. 3.17) is a large HII region surrounding the central star-forming region with a newly born compact open cluster of B, A, and early stars. The whole nebula neighbors a giant molecular cloud located in the southwestern (bottom-right) direction, which partially shades the nebula.

3.4.2.4 Sgr B2

Sgr B2 is a strong HII region in the central region of the galaxy at $l = 1^\circ.2,\ b = 0^\circ$, comprising several compact radio- emitting sources, and is observed to be expanding at an ultrasupersonic velocity (several tens of km s^{-1}). The Sgr B2 HII region is surrounded by a clumpy shell of dense molecular cores each $\sim 10^4\, M_\odot$, and the shell is in touch with the HII region and will be the site of star formation in

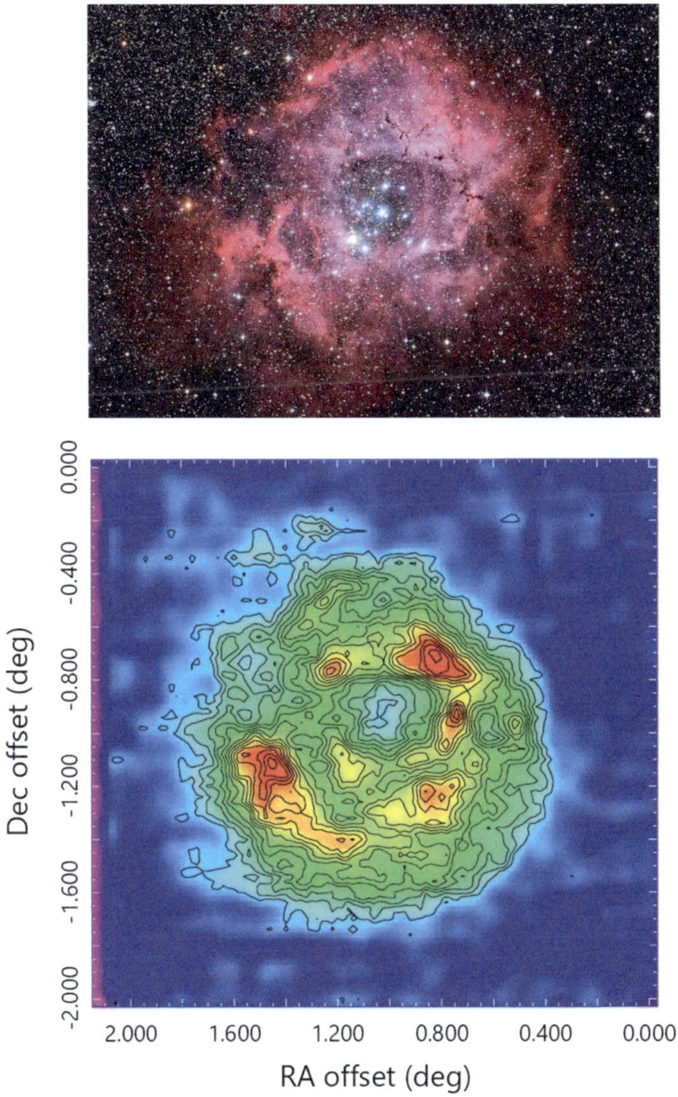

Fig. 3.17 HII region, Rosette Nebula, emitting Hα line red emission (*top*: Photo, Courtesy: Kiso Observatory, The Univ. Tokyo) and thermal radio emission at 10 GHz observed with the Nobeyama 45-m telescope (*bottom*). The nebula surrounds newly born early- type stars in the center. The optical nebula is shadowed by a foreground dark cloud in the bottom-right corner, and the radio nebula is not obscured, showing a fat and rounder shape

the near future. The dense molecular shell may be formed by a shock compression by the expanding HII gas, and the clumps formed by the gravitational fragmentation of such a compressed shell. The whole region is located in the central region of a giant molecular complex of diameter 50 pc and mass $10^6\,M_\odot$.

The star-formation efficiency defined by the ratio of the UV photon intensity necessary for the excitation of the HII gas to the mass of molecular gas (fuel for the star formation) is smaller compared to the SFE for the other regions. This indicates that Sgr B2 will be in an extremely young phase of star formation, embedded in a potential site of much more intense star formation in the future: Sgr B2 region is supposed to be in a "preburst" phase.

3.5 Supernova Remnant

3.5.1 Supernovae (SN) and Supernova Remnants (SNR)

Stars with masses less than the Chandrasekhar limit ($1.4\,M_\odot$) at the final stage of evolution die "quietly" via white dwarfs. If the mass of a star at the final stage of evolution exceeds the limit, which usually occurs for stars with masses of a few $M_\odot$ during the main sequence after which the star loses mass due to stellar wind, it explodes as a supernova. The remnants of the SN expand as strong shock waves through the interstellar gas, and are observed as supernova remnants (SNR) [2, 5, 13, 17].

The rate of SN explosions in the galaxy is estimated to be approximately 1 SN per 30 yr: $r_{\rm SN} \sim 1/30\,{\rm yr}^{-1}$. During the explosion a huge amount of kinetic energy ($E_0 \sim 10^{49} - 10^{51}$) per SN is supplied to the interstellar medium, giving rise to a strong shock wave. The shock wave expands into the interstellar gas in a spherical shell of the shock front, and is called supernova remnants (SNR).

Type I SN: Supernovae as the result of explosion of population II (less massive) stars are classified as type I SN, and occur in the old population disk. Type I SN has the maximum absolute magnitude of about -20 mag., and the rising timescale is as fast as a few days, and the decaying time is also as short as a couple of weeks. The energy released by type I SN is on the order of $10^{49} \sim 10^{50}$ erg, and the kinetic energy given to the ISM is smaller compared to type II SN as described below.

Type II SN: Massive stars of O and B types (population I) explode as type II SN, and are observed mainly in spiral arms. The maximum absolute magnitude of this type of SN is about 18 mag., and the light curve lasts for a month or more. Although the luminosity of type II SN is less than that of type I, they supply a greater amount of kinetic energy to the ISM through their strong shock waves induced by the larger amount of ejecta from the envelopes of the progenitor massive stars.

SNRs are subjects for the physics of gas dynamics and shock wave propagation through the interstellar gas and clouds. SNRs supply a considerable amount of kinetic as well as thermal energy to the interstellar gas, acting to maintain the turbulence in the ISM. SNRs are therefore deeply coupled to the heating and compression of gas clouds, star formation, excitation of turbulence, compression of magnetic fields, and acceleration of cosmic rays, and are also the sources of heavy elements. These will further influence the ISM physics of the galactic disk and halo as well as the evolution of ISM in the galaxy.

3.5.2 Classification of SNR

Shell-type SNR:

Normal observed SNRs are spherical shells with numerous thin filaments (or sheets), and are most naturally taken as expanding spherical shock waves driven by a point explosion at the center. The Cygnus Loop, IC 443, and S147 are the most typical examples of late stages (Fig. 3.18). These SNR have diameters of about 10–30 pc and are expanding at a supersonic velocity of $\sim 100\,\mathrm{km\,s^{-1}}$. Cas A is an example of the young shell type SNR, which is expanding at a velocity as high as $\sim 2000\,\mathrm{km\,s^{-1}}$ (Fig. 3.19). Table 3.2 lists some well-known SNRs [5].

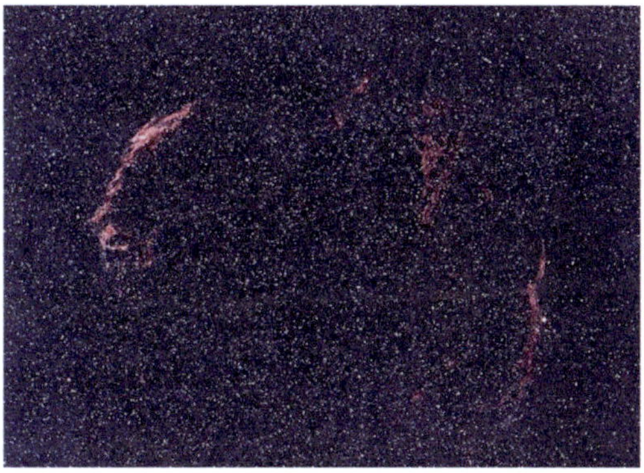

Fig. 3.18 Old supernova remnant Cygnus Loop in optical wavelength (Courtesy: Kiso Observatory)

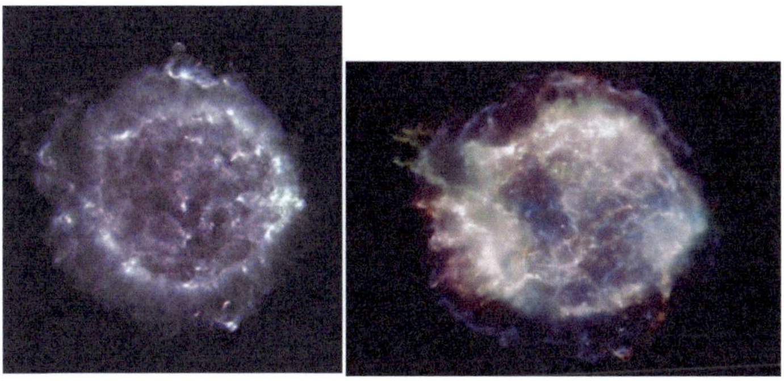

Fig. 3.19 Supernova remnant Cas A in radio (*left*: Image courtesy of NRAO/AUI, http://images.nrao.edu/395) and X rays (*right*: APOD 2002, http://apod.nasa.gov/apod/ap020824.html)

Table 3.2 Typical SNRs and classification

No. (l, b)	A.Dia (′)	Dia. (pc)	Dist. (kpc)	Σ_{1GHz} $Wm^{-2}\,Hz^{-1}Str^{-1}$	Classif.	Remarks
G004.5 + 6.8	3.8	3.3	3	$2.1\ 10^{-19}$	Shell	Kepler SN1604
G006.4–0.1	49	34	2.4	$2.0\ 10^{-20}$	Shell	W28
G023.3–0.3	21	30	5.0	$2.0\ 10^{-20}$	Shell	W41
G074.0–8.6	180	41	0.8	$9.3\ 10^{-20}$	Shell	Cyg. Loop
G111.7–2.1	4.2	3	2.8	$2.9\ 10^{-17}$	Shell	Cas A
G120.1 + 1.4	7.9	12	5.0	$1.4\ 10^{-19}$	Shell	Tycho SN1572
G180.0–1.7	166	79	1.6	$3.0\ 10^{-22}$	Old shell	S147
G184.6–5.8	3.5	2.2	2.2	$1.2\ 10^{-17}$	Filled Cen.	Crab SN1054
G189.1 + 2.9	40	23	1.9	$1.5\ 10^{-20}$	Shell	IC443
G263.9–3.0	256	44	0.6	$4.0\ 10^{-21}$	Amorph	Vella Nebula
G327.6 + 14.0	34	13	1.3	$2.7\ 10^{-21}$	Shell	SN1006

Filaments comprise various components of different temperatures between 10^3 and 10^5 K, emitting various emission lines characteristic of the temperatures. They are, for example, Hα, Hβ ($n_e \sim 100\,cm^{-3}$; $\sim 10^4$ K), HeI, [NII], OI, OIII, [OII], [OIII] ($n_e \sim 10^3 - 10^4\,cm^{-3}$; several 10^4 K), [NeIII], or [SII] lines. The interior of the shell is usually filled by high-temperature ($10^6 \sim 10^7$ K) ionized gas, emitting X-rays.

Shell type SNRs emit synchrotron radio emission from the filaments (sheets), having a steep spectral index ($\alpha \sim -0.5 - 1$). The radio emission is usually polarized in the direction perpendicular to the shell surface, which is because the magnetic field is compressed along the shell surface.

Filled-center SNR:

Many SNRs with amorphous morphology have been found that show no clear shell structure but have a filled-center morphology. These SNRs are called filled-center SNR. Their spectral index is flatter ($\alpha \sim 0$) than that for the shell type, suggesting that younger cosmic-ray electrons are accelerated or supplied more constantly.

The Crab Nebula (Tau A) is a typical SNR of this type, which comprises numerous filaments that are more chaotic and distributed over the remnant. This type of SNR shows no clear shell structure, and the filaments are believed to be excited by injection of high-energy particles from the central neutron stars (pulsar). A number of filled-type SNRs have been discovered in this decade, which were difficult to find in the early days because of their irregular shapes.

3.5.3 $\Sigma - D$ Relation and Distribution of SNR

The surface brightness of a supernova remnant is defined by

$$\Sigma_\nu = F_\nu / \Omega, \tag{3.110}$$

where F is the total flux density of the source at frequency ν and Ω is the apparent solid angle (in steradian) of the remnant. Empirically, the surface brightness is known to be related to the true linear diameter of a remnant D and its height from the galactic plane z as

$$\Sigma_{1\,\mathrm{GHz}} \sim 2.9 \times 10^{-15} D^{-3.9}, \tag{3.111}$$

where $\Sigma_{1\,\mathrm{GHz}}$ is the surface brightness at 1 GHz in W m^{-2} Hz^{-1} str^{-1}, and D and z are in pc [12]. This is called the $\Sigma - D$ relation, and is used to obtain the distance to a SNR by measuring the total flux density and apparent extent (Fig. 3.20).

Given the diameter of a SNR, its distance d can be estimated by $d = D/\theta$, where θ is the angular diameter of the SNR. From d and the position in the sky (l, b) we can derive the three-dimensional position of the SNR.

From the almost complete sampling in the solar vicinity, the total number of SNRs in the galaxy is estimated to be on the order of $25 \times (\pi 10\ \mathrm{kpc}^2/\pi 1\ \mathrm{kpc}^2) \sim 2500$. Because the mean lifetime of a SNR is roughly 10^5 yr, the rate of SN explosions can be estimated as $r_{\mathrm{SN}} \sim 2500/10^5 = 1$ per 40 yr. (From a more accurate study we know that the rate is about 1 per 20–30 yr.)

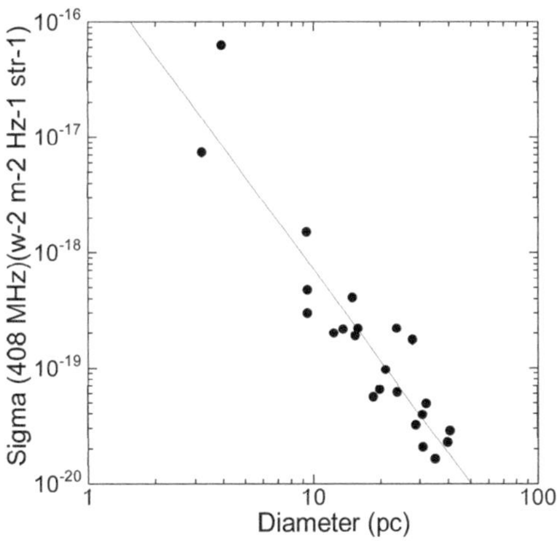

Fig. 3.20 Surface brightness-diameter ($\Sigma - D$) relation for SNRs at 408 MH [12]

3.5.4 Evolution of a SNR

The evolution of a spherical shell produced by a SN explosion can be approximated by a spherical shock propagation in a uniform medium (Figs. 3.21 and 3.22). If the shell evolves in an adiabatic condition, namely if the radiative cooling time of the shell is short enough compared to the timescale of the expansion, the propagation of a spherical shock wave produced by a point explosion in a uniform medium can be described by Sedov's similarity solution. Let the thermal + kinetic energy released by a SN explosion be E_0, which is transformed to the kinetic energy of expansion of the surrounding gas of density ρ_0. Suppose that the gas is shock compressed in a thin shell of radius r which is expanding at a velocity v. Then we have

$$E_0 \sim \frac{1}{2}\left(\frac{4\pi}{3}r^3\rho_0\right)v^2, \tag{3.112}$$

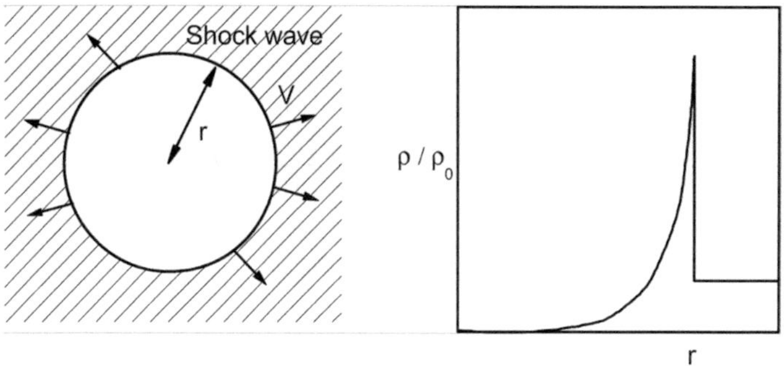

Fig. 3.21 A spherical shocked shell produced by a point explosion in a uniform ISM by a SN

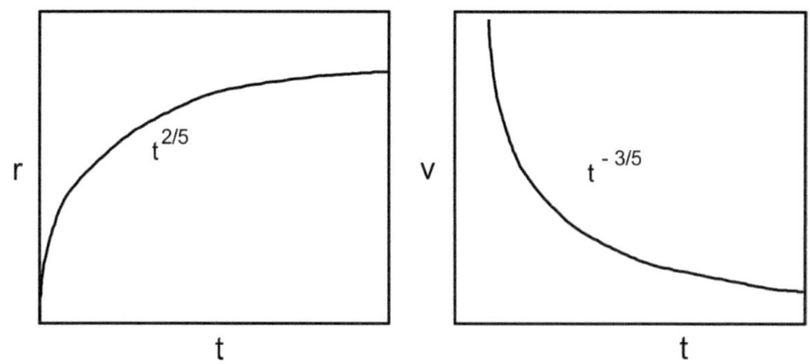

Fig. 3.22 Time variation of the radius and expansion velocity of an adiabatic shell of a spherical shock wave

where $M = (4\pi/3)r^3\rho_0$ is the accumulated (snow-plowed) mass of the ISM in the shocked shell. Because $v = dr/dt$, we can solve the above equation to yield

$$v \sim \left(\frac{E_0}{\rho_0}\right)^{1/5} t^{-3/5}, \tag{3.113}$$

and the radius increases with t as

$$r \sim \left(\frac{E_0}{\rho_0}\right)^{1/5} t^{2/5}. \tag{3.114}$$

The energy given to the surrounding ISM is about $E_0 \simeq 10^{50-51}$ erg for type II SN and 10^{49-50} erg for type I.

These formulations only apply to a limited phase of the evolution of a SNR when the compressed gas in the shell is approximated as adiabatic. In reality, however, a SNR experiences various phases according to the time elapsed from the explosion as well as depending on the physical condition of the surrounding ISM. A typical evolution of a SNR is described as follows (Fig. 3.23).

3.5.4.1 Free Expansion Phase ($t \sim 0$–10^3 yr)

At the very initial phase of expansion, when the mass of shock-accumulated gas from the ISM is smaller than the mass of ejecta from the SN, the ejecta expand almost freely at a velocity of v_0, which is related to the energy as $E_0 \sim (1/2)v_0^2$. For a type II SN, the mass and explosive energy are about $M_0 \sim 5\,M_\odot$ and $E_0 \sim 10^{51}$ erg, respectively, and therefore, $v_0 \sim 5000$ km s^{-1}; For a type I SN, $M_0 \sim 1\,M_\odot$ and $E_0 \sim 10^{49}$ erg, and $v_0 \sim 1000$ km s^{-1}. The Crab Nebula is an example of this phase.

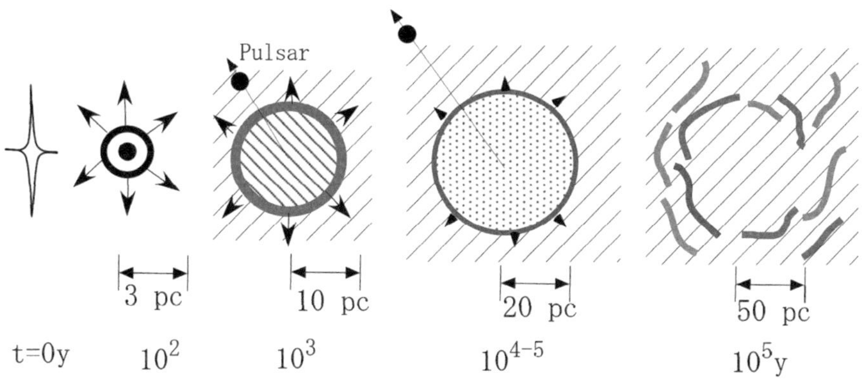

Fig. 3.23 Evolution of a SNR

3.5.4.2 Adiabatic Shocked-Shell Phase ($t \sim 10^3$–10^4 yr)

When the accumulated mass from the ISM ($M \sim (4\pi/3)r^3\rho_0$) begins to exceed the initial mass of the ejecta(M_0), the shell expansion begins to suffer deceleration, and a shocked shell forms. This occurs at about $t \sim 10^3$ yr and $r \sim 3$ pc. If the cooling timescale of the shell is smaller than the age (timescale of the expansion) $t_{\rm cool} < t_{\rm exp} \sim r/(dr/dt) \sim r/v$, the shell can be approximated by an adiabatic spherical shock wave as described above. The cooling time for a shocked gas of density $n \sim 4n_0$ with $n_0 \sim 1\,{\rm H\,cm^{-3}}$ and temperature 10^6 K corresponding to an expansion velocity $100\,{\rm km\,s^{-1}}$ is about 10^4 yr. Therefore, the adiabatic expansion applies for SNR of $t < 10^4$ yr.

3.5.4.3 Dense Shell Formation ($t \sim 10^4 - 10^5$ yr)

At this stage the radiative cooling of shock-compressed gas cannot be ignored, because the cooling time becomes comparable to or shorter than the expansion timescale of the remnant. The adiabatic approximation no longer applies, and the shell is more effectively compressed by the cooling. If the cooling is effective enough, the shock compression amounts to $M^2 \sim (v_{\rm shell}/c_{\rm s})^2$, where M is the Mach number and $c_{\rm s}$ is the sound velocity. The shell is fragmented into numerous sheets and filaments by thermal as well as hydrodynamical instabilities. The shock-compressed filaments and sheets radiate line emissions and cool down to denser fragments with density 10^2–$10^3\,{\rm cm^{-3}}$. The thickness of the filaments is on the order of 0.1 pc, much smaller than the SNR radius, 20–30 pc. Typical examples are seen in the Cygnus Loop, IC 443, and S147.

3.5.4.4 Late Phase; Fading into ISM

When the shell radius exceeds ~30 pc at $t > 10^5$ yr, the expansion velocity decreases to a few tens of ${\rm km\,s^{-1}}$. The shell suffers from significant deformation due to the interaction with turbulent ISM and clouds. When the shell expands further and is decelerated, with the expansion velocity becoming comparable to the sound velocity of the ISM, the SNR fades into the ISM as sound waves. Any SNR disappears at this stage, finally reducing the kinetic energy to the turbulent motion of the ISM.

3.5.5 *Interaction with the ISM*

Because of interaction with interstellar clouds the SNR shell deforms from its spherical shape (Fig. 3.24). When it encounters a cloud of similar size to the shell radius, the SNR will attain a highly asymmetric shape: the Cygnus Loop and IC 443

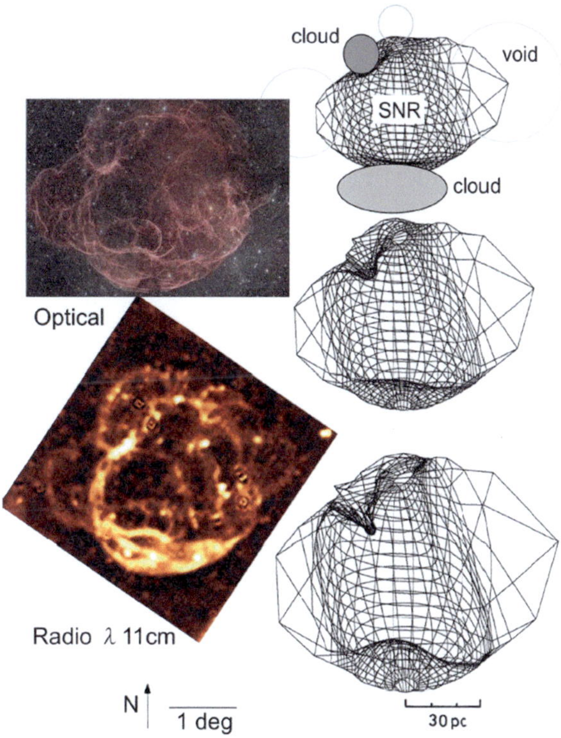

Fig. 3.24 Interaction of a SNR shell with ISM with clouds and voids, compared with an Hα photo of the SNR, S147 (APOD 2011, http://apod.nasa.gov/apod/ap110212.html (Courtesy by Dr. Nobuhiko Miki)); Radio image at $\lambda = 11$ cm [19]

are typical examples of such interaction. If the size and mass of the cloud are large enough, the propagation of a shock wave is strongly refracted, and a focusing shock into the cloud often occurs. The focusing shock compresses the cloud, resulting in an implosion of the cloud. On the other hand, a small cloud is often "evaporated" by the shock heating.

Even if the surrounding ISM is uniform, the surface of the SNR shell is often fragmented due to hydrodynamical instabilities. As the Sedov solution indicated, the SNR shell is decelerated by

$$d^2r/dt^2 \sim -t^{-8/5}. \tag{3.115}$$

In a coordinate system in which the shell surface is at rest, the acceleration points toward the outer region: the SNR shell lies on the ISM with the acceleration (apparent gravity) pointing from a denser region to a rare region. Therefore, a Rayleigh-Taylor instability grows. This instability is the major cause of the filamentary structures observed in SNRs, and also for the clumpy structures observed in such early phase (and therefore, stronger deceleration) SNRs as the Cas A. According to the compression of the gas as well as the interstellar magnetic fields, cosmic rays are accelerated, and such filaments and clumps are strong radio-emitting regions.

3.5.6 Implications of SNRs for Galaxy Evolution

3.5.6.1 SNR and Star Formation

The lifetimes of massive stars (O stars) are approximately 10^7 yr (1.2×10^7 yr for $15\,M_\odot$; 3×10^7 yr for $10\,M_\odot$; 9×10^7 yr for $5\,M_\odot$). Therefore, type II SNs explode in the vicinity of the places where the progenitor stars were born such as near the SF region rich in MC and GMCs. The encounter of a SNR shock wave with a molecular cloud would, then, further compress it to create another SF region. Hence, a SN explosion followed by a shock wave (e.g., SNR) is another type of triggering mechanism of the sequential SF in the cloud complex.

It also happens that large-scale shock waves produced by SNR hit more distant MCs and GMCs. This would further trigger SF inside the clouds, leading to a larger-scale "intercloud" sequential SF. In this way, the SF propagates from clouds to clouds, and the intercloud sequential SF acts as a stochastic formation of spiral patterns of OB stars in a galactic disk in differential rotation.

3.5.6.2 Energy Supply to ISM Turbulence

If there is no energy supply, the interstellar turbulence (random motion of interstellar clouds) will decay within 10^7 yr ($\sim$ collision time among the clouds) because of dissipation through elastic collisions among clouds. The kinetic energy necessary for maintaining the turbulence is thought to be supplied by kinetic energy from SN explosions. Because the total mass of ISM is $M_{\rm ISM} \sim 0.05 M_{\rm G} \sim 10^{10} M_\odot$, the kinetic energy of turbulent motion is

$$E_{\rm turb} \sim (1/2) M_{\rm ISM} v_{\rm t}^2 \sim 5 \times 10^{54}\,{\rm erg}. \tag{3.116}$$

The energy dissipation rate is on the order of

$$dE_{\rm t}/dt \sim E_{\rm t}/\tau \sim 5 \times 10^{47}\,{\rm erg/yr}. \tag{3.117}$$

If $\sim$10% of the total energy released by SN explosions is finally given to the ISM turbulence as kinetic energy, the rate is estimated to be

$$dE/dt \sim 0.1 \times E_0 r_{\rm SN} \sim 3 \times 10^{48}\,{\rm erg/yr}, \tag{3.118}$$

where

$$r_{\rm SN} \sim 1/(30\,{\rm yr}) \tag{3.119}$$

is the rate of type II SN explosions in the galaxy. This is a much larger amount compared to the dissipation rate of the ISM turbulent energy; thus only a fraction of the energy released by SN explosions is sufficient for a steady excitation of the ISM turbulence.

3.5.6.3 Energy Supply to the Halo Gas and Galactic Winds

The rest of the kinetic energy from SN would be used for heating the intercloud gas to high temperatures of $\sim 10^6 - 10^7$ K as well as for accelerating the heated gas flow into the halo. As has been shown, the scale thickness of the SNR distribution is about $z_{\rm SNR} \sim 60$ pc, which indicates that about 10% of SN explosions occurred in the space high above the galactic plane at $|z| > 100$ pc. At such heights, about half the explosive energy is used to accelerate the gas into the halo (Fig. 3.25). In order for a gas to be blown into the halo, the shell speed must be greater than $100\,{\rm km\,s^{-1}}$, which is realized if the shell radius is smaller than 20 pc and the accumulated shell mass $200\,M_\odot$, assuming an efficiency of 10% for E_0 to kinetic energy. The total mass blown into the halo and the kinetic energy supplied to the halo gas are then estimated as

$$dM_{\rm gas\ to\ halo}/dt \sim 0.1 \times 200\,M_\odot \times 1/30\,{\rm yr}^{-1} \sim 0.7\,M_\odot\,{\rm yr}^{-1}, \tag{3.120}$$

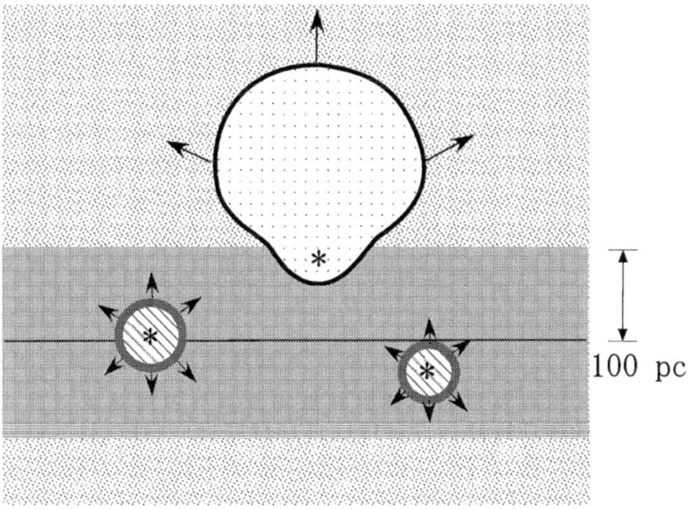

Fig. 3.25 Shock wave caused by a SN exploded at high z expands into the halo

and

$$dE_{\text{to halo}}/dt \sim 1.3 \times 10^{47}\ \text{erg yr}^{-1}, \tag{3.121}$$

respectively.

3.5.6.4 Metal Abundance

The primeval galaxy was composed of H, He, and only a small amount of heavy elements after the Big Bang at a ratio in a mass of

$$X : Y : Z = 0.75 : 0.25 : 0.001. \tag{3.122}$$

On the other hand, the present heavy element ratio is about

$$Z \simeq 0.02 - 0.05 \tag{3.123}$$

in the solar vicinity. Therefore, the metal abundance has increased by one to two orders of magnitude in the last 15 billion years since the formation of the galaxy. This "pollution" of the ISM occurred partly because of the stellar wind from massive stars and partly due to heavy element dissipation from the interiors of massive stars due to SN explosion.

Suppose that a single SN explosion distributes heavy elements of mass of $0.1\,M_{\odot}$ into interstellar space. Then the total amount of heavy elements of the ISM in the galaxy increases at a rate

$$dM_{\text{Z}}/dt \sim 0.1 M_{\text{SN}}/20\,\text{yr}. \tag{3.124}$$

The rate of increase in the Z abundance is given by

$$dZ/dt \sim (dM_{\text{Z}}/dt)/M_{\text{ISM}}. \tag{3.125}$$

Because $M_{\text{ISM}} \sim 10^{10}\,M_{\odot}$, we have

$$dZ/dt \sim 1.7 \times 10^{-12}, \tag{3.126}$$

or

$$Z \sim 1.7 \times 10^{-12} t \tag{3.127}$$

for $M_{\text{SN}} \sim 5\,M_{\odot}$. For the age of the galaxy $t \sim 1.5 \times 10^{10}$ yr, we have the present Z to be $Z \sim 0.03$. The sun was born 4.6×10^{9} yr ago or at $t = 10^{10}$ yr, at which time the metal abundance was $Z \sim 0.02$. This is approximately equal to the solar value ($Z = 0.017$).

References

1. Chandrasekhar, S.: Hydrodynamic and Hydromagnetic Stability. Oxford University Press, London, Chap. X, XI. (1968) [Instabilities]
2. Chevalier, R.A.: The interaction of supernovae with the interstellar medium. ARAA **15**, 175 (1977). [SNR; ISM]
3. Elmegreen, B., Lada, C.: Sequential formation of subgroups in OB associations. ApJ **214**, 725 (1977). [SF propagation]
4. Field, G.B.: Thermal instability. ApJ **142**, 531 (1965). [Thermal instability]
5. Green, D.A.: A catalogue of 294 Galactic supernova remnants. Bull. Astron. Soc. India **42**, 47 (2014). [SNR catalogue]
6. Hou L.G., Han J.L.: The observed spiral structure of the Milky Way. A&A **569**, A125 (2014).
7. Kennicutt, Jr. R.C.: The star formation law in galactic disks. ApJ **344**, 685 (1989). [Star formation in galaxies]
8. Larson, R.B.: Effects of supernovae on the early evolution of galaxies. MNRAS **169**, 229 (1974). [Stellar mass function]
9. McKee, C.F., Hollenbach, D.J.: Interstellar shock waves. ARAA **18**, 219 (1980). [Interstellar shock]
10. Nakanishi, H., Sofue, Y.: Three-dimensional distribution of the ISM in the Milky Way Galaxy: II. The molecular gas disk. PASJ **58**, 847 (2006)
11. Nakanishi, H., Sofue, Y.: Three-dimensional distribution of the ISM in the Milky Way galaxy. III. The total neutral gas disk. PASJ **68**, 5 (2016)
12. Pavlović, M.Z., Urošević, D., Vukotić, B., et al.: The radio surface-brightness-to-diameter relation for galactic supernova remnants: sample selection and robust analysis with various fitting offsets. ApJS **204**, 4 (2013). [Sigma-D relation of SNR]
13. Raymond, J.C.: Observations of supernova remnants. ARAA **22**, 75 (1984). [SNRs]
14. Scalo, J.M.: The stellar initial mass function. Fund. Cosmic Phys. **11**, 1 (1986). [Initial mass function]
15. Shu, F.H., Adams, F.C., Lizano, S.: Star formation in molecular clouds – observation and theory. ARAA **25**, 23 (1987). [Star formation]
16. Shull, J.M., Hollenbach, D.J.: H_2 cooling, dissociation, and infrared emission in shocked molecular clouds. ApJ **220**, 525 (1978). [Cooling function]
17. Weiler, K.W., Sramek, R.A.: Supernovae and supernova remnants. ARAAA **26**, 295 (1988). [SN, SNR]
18. Woodward, P.R.: Theoretical models of star formation. ARAA **16**, 555 (1978). [ISM shock; cloud instability]
19. Xiao, L., Fuerst, E., Reich, W., Han, J.L.: Radio spectral properties and the magnetic field of the SNR S147. AA **482**, 782 (2008). [Radio image of SNR S147]

Chapter 4
Galactic Structure

4.1 The Milky Way Galaxy

4.1.1 Edge-On View of the Galaxy

The Milky Way in the sky is an edge-on view of our galaxy, a typical spiral galaxy of Sb type of Hubble classification. It is a projection of stars and ISM in the galaxy. There have been a number of all-sky surveys and galactic-plane surveys at various frequencies. Figures 4.1, 4.2, 4.3, and 4.4 show compilations of multiwavelength all-sky maps of the galactic disk in HI and CO line emissions, FIR (far infrared) emission, and in the radio continuum at 408 MHz.

4.1.2 Distance to the Galactic Center, R_0

The distance to the galactic center from the Sun (R_0) is the most fundamental parameter describing the galactic structure. The distance is related to such fundamental quantities as the mass and size of the galaxy and those of individual objects, and thus their luminosities. The distance to the galactic center has been determined by various methods as follows, and the distance widely adopted today is $R_0 \simeq 8$ kpc.

4.1.2.1 Center of Globular Cluster Distribution

In the classical method, the center of the distribution of globular clusters is assumed to coincide with the galactic center (Fig. 4.5). Distances to globular clusters are measured by observing RR Lyr stars by applying the period-luminosity relation. Distances to clusters that do not involve RR Lyr stars are measured by the spectroscopic parallax method using Hertzsprung-Russell (HI) diagrams.

The original version of this chapter was published without printing chapter abstract. The abstracts for each chapter are updated in the front matter in the current version of the book. An erratum to this chapter can be found at https://doi.org/10.1007/978-981-10-3445-9_7.

Y. Sofue, *Galactic Radio Astronomy*, Lecture Notes in Physics 935,
DOI 10.1007/978-981-10-3445-9_4

Fig. 4.1 Edge-on views of the galaxy in radio, FIR, optical emission, and in X-rays (Courtesy of NASA, http://mwmw.gsfc.nasa.gov/mmw_product.html)

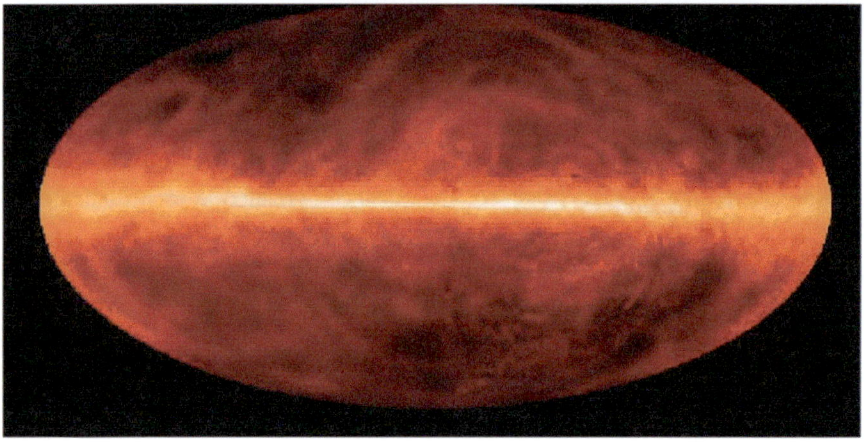

Fig. 4.2 Edge-on view of the galaxy in HI by integrated 21-cm line intensity [7]

From the distances and their positions in the sky, three-dimensional positions of the clusters, and thereby the distribution in the galaxy, are obtained. The galactic globular clusters are known to be distributed almost spherically in the galactic bulge and halo. Assuming that the center of this spherical distribution coincides with the galactic center, the distance R_0 has been obtained to be about 8 to 10 kpc.

Fig. 4.3 Edge-on view of the galaxy in CO integrated 2.6-mm line intensity, showing molecular gas (*top*) [3], and in FIR at 100 μm (dust) (*bottom*: http://apod.nasa.gov/apod/ap000517.html)

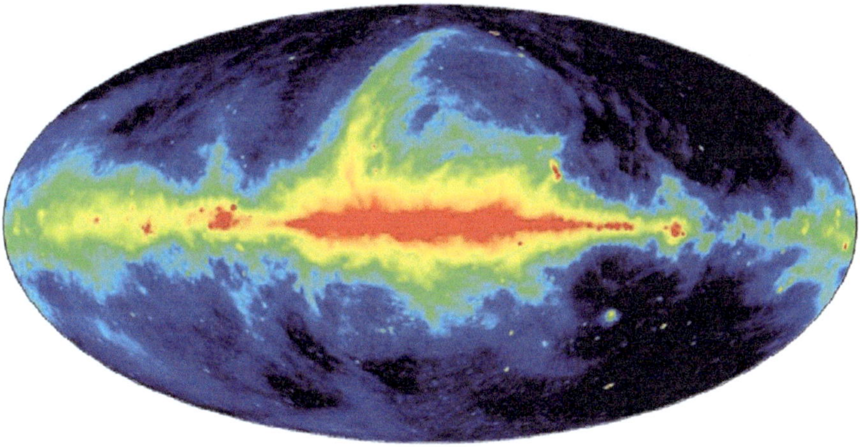

Fig. 4.4 Edge-on view of the galaxy in radio continuum at 408 MHz [4]

4.1.2.2 Bulge Center by Infrared Variable Stars

The central bulge of the Milky Way exhibits a round shape whose center is supposed to coincide with the galactic center. Cepheid variables are the most reliable distance indicator of stellar objects, but the heavy extinction in the optical wavelengths makes it difficult to apply the method to the galactic center direction. Long-period variable

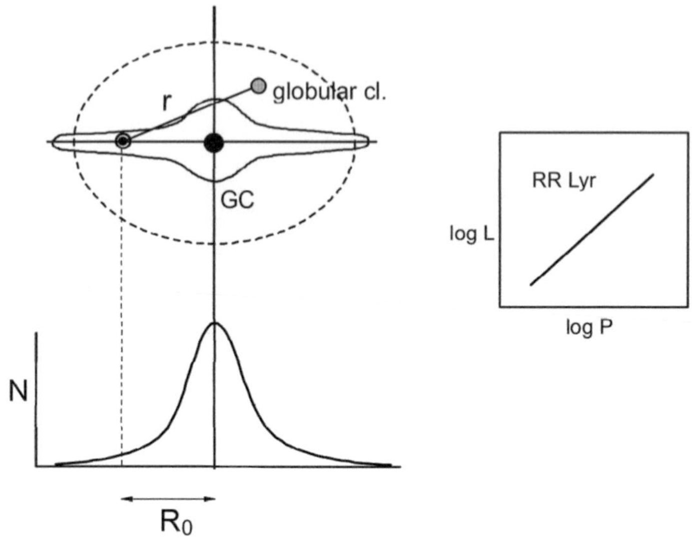

Fig. 4.5 The distribution of globular clusters around the galaxy, and the distance to the galactic center from the Sun. The inset diagram illustrates the period-luminosity relation for RR Lyr stars in globular clusters

stars in infrared wavelengths such as Mira variables are extensively used in order to avoid this difficulty. According to distance measurement applying the period-luminosity relation to the variables, the center of a supposed spheroidal distribution, the stars have been determined to be 8.24 ± 0.42 kpc [8].

4.1.2.3 VLBI Parallax of Sgr A*

The nucleus of our galaxy is the compact radio source Sgr A* associated with a massive black hole. For its compactness and high radio continuum brightness, Sgr A* is a good target for parallax measurement by very long baseline interferometer (VLBI). Its paralactic motion can also be used to measure the solar rotation speed by using background quasars as reference positions. Extensive parallax observations have been obtained using VLBI and very long baseline array (VLBA), and the parallax has been measured to be $p \simeq 0.13$ mas, corresponding to a distance of 8 kpc.

4.1.2.4 Moving-Cluster Method Using Maser Sources

The moving cluster method is a basic method to determine the distance of a cluster of stars, and has been applied to nearby open clusters such as the Hyades cluster to derive its distance and HR diagrams of the member stars. This method is applicable to the galactic center (GC) clusters of maser sources, and the principle can be applied to VLBI radio line measurements.

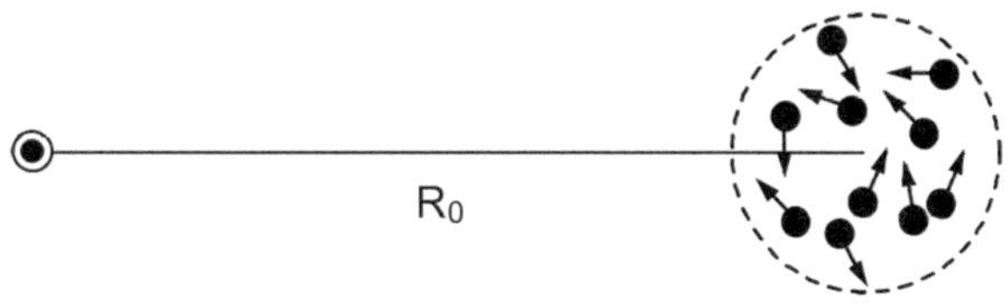

Fig. 4.6 Radial velocity dispersion and a dispersion of proper motions in a spherical cloud composed of many maser-line sources give us the distance

Consider a spherical cloud near the galactic center, in which many H_2O maser sources are distributed spherically and with random internal motion (Fig. 4.6). Let the angular velocity (proper motion) of a cloud be $\mu = d\theta/dt$, the velocity dispersion perpendicular to the line of sight σ_p, and the velocity dispersion of the maser sources in the line of sight σ_r. If angular velocities of many maser sources are measured by VLBI observations and a dispersion of their proper motions with respect to the cloud center is obtained, we have

$$\sigma_p \simeq R_0 \langle \mu^2 \rangle^{1/2} \simeq R_0 \left(\frac{\Sigma \mu^2}{N-1} \right)^{1/2}. \tag{4.1}$$

On the other hand, σ_r is obtained by measuring the dispersion of radial velocities of the sources from maser-line observations. From the assumption of a spherical system, we have

$$\sigma_r \simeq \sigma_p, \tag{4.2}$$

and therefore,

$$R_0 \simeq \frac{\sigma_r}{\langle \mu^2 \rangle^{1/2}}. \tag{4.3}$$

Here, the quantities on the right-hand side of the equation are observables from the VLBI line observations. The galactic center distance determined by this method is $R_0 \simeq 8$ kpc.

4.1.2.5 X-Ray Bursters

When the luminosity of a star exceeds a threshold value, the envelope of the star is blown away by its own radiation pressure overcoming the stellar gravity. This threshold (critical) luminosity is called the Eddington limit, and is uniquely determined by the mass of a star. X-ray busters are known to be extremely luminous at their bursting phase. However, this maximum luminosity should not exceed the Eddington limit. If we assume that the bursting-phase luminosity is approximately the Eddington luminosity, the distance can be estimated by comparison with the apparent magnitude. By applying this method to X-ray busters in the galactic bulge, a distance of $R_0 \sim 7$ kpc has been obtained.

4.1.3 Rotation of the Sun Around the Galactic Center

4.1.3.1 Oort's Constants, *A* and *B*

Suppose that the Sun is rotating around the galactic center on a circular orbit of radius R_0 at a velocity V_0 (Fig. 4.7). A galactic object is located in the direction l ($b \sim 0°$) at a distance r from the Sun, which is also rotating around the galactic center at a velocity V with radius R.

The radial velocity of the object as seen from the sun is given by

$$v_\mathrm{r} = V \cos\ \alpha - V_0 \sin\ l, \tag{4.4}$$

where α is the angle between the velocity vector of the object and the line of sight. Knowing that

$$\frac{\sin\ l}{R} = \frac{\cos\ \alpha}{R_0} \tag{4.5}$$

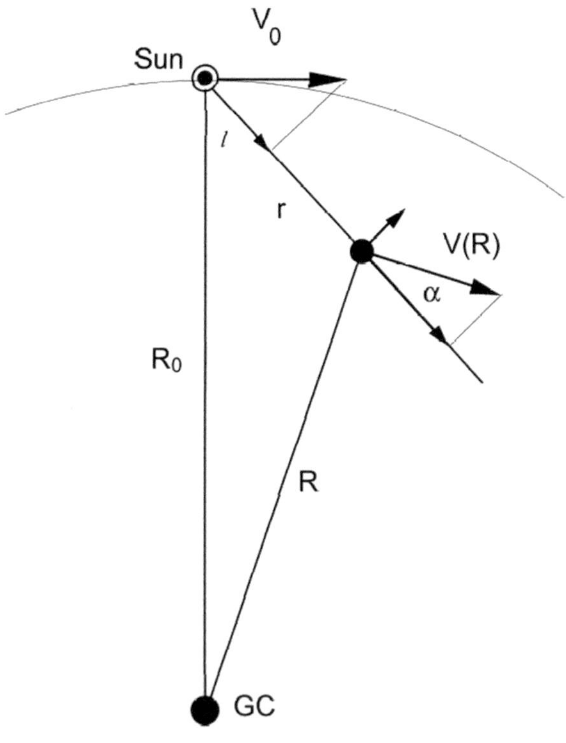

Fig. 4.7 The Sun and a galactic object are rotating around the galactic center on circular orbits

we have

$$v_{\rm r} = \left(V \frac{R_0}{R} - V_0 \right) \sin\ l. \tag{4.6}$$

Introducing angular velocities $V = R\omega$ and $V_0 = R_0\omega_0$, we can rewrite the equation as

$$v_{\rm r} = (\omega - \omega_0) R_0 \sin\ l. \tag{4.7}$$

The velocity of the object perpendicular to the line of sight is written as

$$v_{\rm p} = \frac{r}{R}(R_0 \cos\ l - r) - V_0 \cos\ l, \tag{4.8}$$

which is rewritten as

$$v_{\rm p} = (\omega - \omega_0) R_0 \cos\ l - \omega r. \tag{4.9}$$

In the solar vicinity, $r \ll R_0, R$, we have

$$\omega - \omega_0 = \left[\frac{d\omega}{dR} \right]_{R=R_0} (R - R_0), \tag{4.10}$$

and

$$\frac{d\omega}{dR} = \frac{d}{dR} \left(\frac{V}{R} \right) = \frac{1}{R} \frac{dV}{dR} - \frac{V}{R^2}. \tag{4.11}$$

Therefore, the radial velocity can be written as

$$v_{\rm r} = \left[\frac{dV}{dR} - \frac{V_0}{R_0} \right] (R - R_0)\ \sin\ l. \tag{4.12}$$

Because

$$R - R_0 \simeq r \cos\ l, \tag{4.13}$$

we have

$$v_{\rm r} = \left[\frac{dV}{dR} - \frac{V_0}{R_0} \right] \frac{r\ \sin\ 2l}{2}. \tag{4.14}$$

Thus we obtain

$$v_{\rm r} = A\ r\ \sin\ 2l \tag{4.15}$$

with

$$A = \frac{1}{2}\left[\frac{V_0}{R_0} - \frac{dV}{dR}\right]_{R=R_0}. \tag{4.16}$$

Here, A is called Oort's A constant. The perpendicular component is given by

$$v_\mathrm{p} = \left[\frac{dV}{dR} - \frac{V_0}{R_0}\right](R - R_0)\ \cos\ l - \omega_0 r$$

$$= \frac{1}{2}\left[\frac{V_0}{R_0} - \frac{dV}{dR}\right] r\ \cos\ 2l - \frac{1}{2}\left[\frac{V_0}{R_0} + \frac{dV}{dR}\right] r. \tag{4.17}$$

Or we can rewrite this as

$$v_\mathrm{p} = (A\ \cos\ 2l + B)\, r \tag{4.18}$$

with

$$B = -\frac{1}{2}\left[\frac{V_0}{R_0} + \frac{dV}{dR}\right] \tag{4.19}$$

called Oort's B constant. Introducing proper motion $\mu = d\theta/dt$ and $v_\mathrm{r} = \mu r$, we rewrite the above expression as

$$\mu = \frac{A\ \cos\ 2l + B}{4.74}\ \ ('' \mathrm{yr}^{-1}). \tag{4.20}$$

Here, Oort's constants A and B are measured in km s^{-1} kpc^{-1}. By observing many stars (objects) with known distance r, proper motion μ, and radial velocity v_r, we can determine A and B by the least-square fitting using Eqs. (4.15) and (4.20). The observations have given

$$A = 15\ \mathrm{km\ s^{-1}\ kpc^{-1}}, \tag{4.21}$$

and

$$B = -10\ \mathrm{km\ s^{-1}\ kpc^{-1}}. \tag{4.22}$$

4.1.3.2 Solar Rotation Velocity, V_0

Equations (4.16) and (4.19) are used to obtain the rotation velocity of the Sun V_0:

$$V_0 = (A - B)R_0, \tag{4.23}$$

and the differential rotation as

$$\frac{dV}{dR}_{(R=R_0)} = -(A+B). \tag{4.24}$$

Because we know the distance to the galactic center as $R_0 = 8\,\text{kpc}$, and $A = 15\,\text{km s}^{-1}\,\text{kpc}^{-1}$ and $B = -10\,\text{km s}^{-1}\,\text{kpc}^{-1}$, we obtain the solar rotation velocity as

$$V_0 = 200\,\text{km s}^{-1}. \tag{4.25}$$

Recent high-accuracy VLBI observations have made it possible to measure the galactic center distance and the solar rotation velocity in more direct ways, trigonometrically measuring parallax, proper motion, and radial velocity of maser sources. VERA (VLBI experiments of radio astrometry) observations have shown an accurate set of the solar constants of $(R_0,\ V_0) = (8.05 \pm 0.45\,\text{kpc},\ 238 \pm 14\,\text{km s}^{-1})$ [6].

4.1.4 *Galactic Rotation Curve*

4.1.4.1 Rotation Curve Within the Solar Circle

We have obtained the distance to the Sun from the galactic center R_0 and rotation velocity V_0. This gives us the opportunity to derive the rotation velocity $V(R)$ of the galactic disk as a function of galactocentric distance R, which is called the rotation curve. We here assume that $V(R)$ varies smoothly with R. We then observe HI and/or CO line profiles of the interstellar gas distributed in the galactic plane toward l ($b \sim 0°$). In the first quadrant of the disk ($0 < l < 90°$) all the gas within the solar circle has positive radial velocity, and the gas outside the solar circle has negative velocity because of the motion of the Sun. A maximum velocity is observed coinciding with a position tangential to the line of sight, and we call this radial velocity the terminal velocity $v_{\text{r max}}$ (Fig. 4.8).

Given this terminal velocity from observation, the rotation velocity $V(R)$ at this position $R = R_0 \sin\ l$ is simply derived by

$$V(R) = v_{\text{r max}} + V_0\ \sin\ l. \tag{4.26}$$

By observing $v_{\text{r max}}$ at various longitudes l, and therefore at various R, we obtain the rotation curve $V(R)$ of the disk in the first quadrant. For the third quadrant ($270° < l < 360°$), the terminal velocity must be negative, and hence, the rotation curve is given by

$$V(R) = -v_{\text{r min}} - V_0\ \sin\ l. \tag{4.27}$$

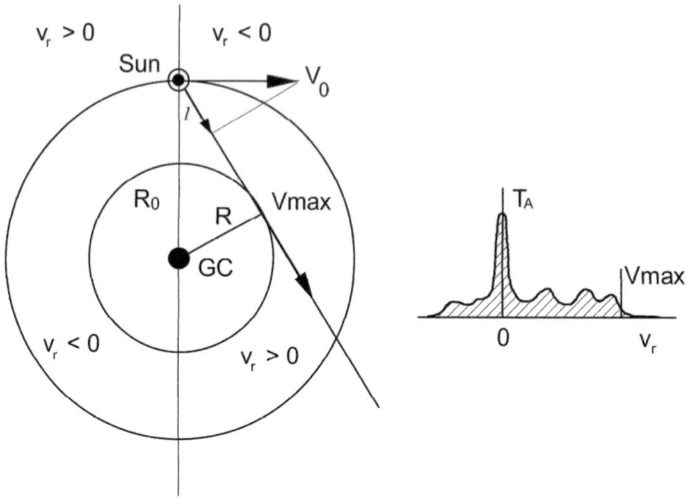

Fig. 4.8 The terminal velocity is simply related to the rotation velocity of the disk within the solar circle

4.1.4.2 Rotation Curve Beyond the Solar Circle

The determination of the rotation curve beyond the solar circle ($R > R_0$) is not straightforward. Sources with *known distances* and their radial velocities are required. Let the distance to an object at (l, b) be r, then the galactocentric distance R is uniquely determined by (l, b, r, R_0). The rotation velocity and an observed radial velocity are related by

$$V(R) = \frac{v_r + V_0 \cos(l - 90°)}{\cos(l + \theta - 90°)}, \tag{4.28}$$

where θ is the galactocentric longitude of the source. The rotation curve has been obtained by observing HII regions, HI clouds, and/or molecular clouds *associated with stars* whose distances have been measured independently by spectroscopic methods (Fig. 4.9).

4.2 Rotation Curve of the Milky Way

The Milky Way Galaxy has been extensively observed in HI, CO, and recombination lines (HII regions), and a number of longitude-velocity ($l - V_{lsr}$) diagrams have been published (e.g., Figs. 4.13 and 4.14). These diagrams have been used to obtain a rotation curve of the galaxy.

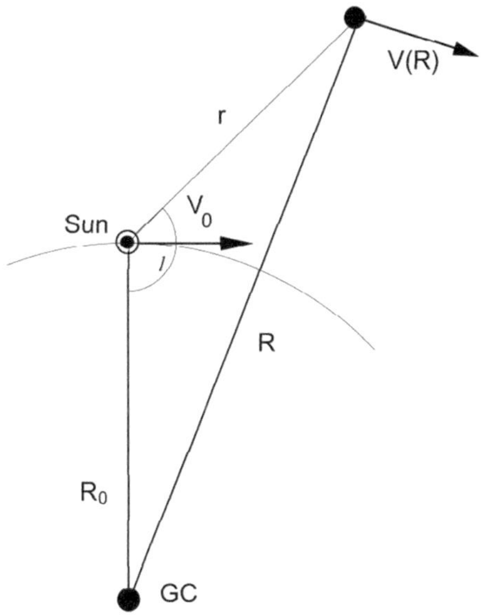

Fig. 4.9 The rotation curve for the outer galaxy is obtained by observing radial velocities of sources with known distance *r*

The central 1° region has been observed mainly in the CO line emissions at a resolution of 2′ (5 pc). Detailed position-velocity (PV) diagrams from the Bell Telephone Laboratory data have been used to derive a rotation curve in the central 150 pc. Figure 4.10 shows a ^{13}CO($J = 2-1$)-line PV diagram averaged in $b = -17'$ to $12'$ (47 pc width). The diagram shows several rigid-body-like ridges, and are identified with rotating ring and arcs of about 120-pc ring. We use these ridges to obtain possible rotation curves. Note that the two symmetrical high-velocity arcs at $V_{\rm LSR} \sim \pm 150$ to $200\,{\rm km\,s^{-1}}$ are parts of the expanding molecular ring, and do not indicate the rotation of the gas. Moreover, these expanding features make up only a few percent of the total emission of the displayed region.

Figure 4.11 shows a rotation curve of the inner 100 pc of the galaxy as obtained from PV diagrams in the inner regions including Fig. 4.10. By combining this rotation curve with that given for the outer region, the most accurate rotation curve of the galaxy has been obtained, as shown in Fig. 4.12.

The rotation curve of the whole galactic disk has been obtained by observations of HII regions in optical H_α and radio recombination lines, HI gas in the 21-cm line emission, and molecular gas by the CO line emission. The rotation curve is known to be almost flat from the central region at $R \sim 200$ pc to the outer regions beyond $R \sim 20$–30 kpc.

Beyond the galactic disk, satellite galaxies and globular clusters are used to estimate approximate circular velocities. A rotation curve covering the widest region of the galaxy is shown in Fig. 4.12 in a logarithmic presentation, which enables us to close up the innermost region as well as to overview the outermost region. The rotation curve is a powerful tool to calculate the mass distribution of the galaxy (Sect. 4.8.1). The velocity remains almost flat until $R \sim 50$ kpc, and then gradually declines until $\sim 150\,{\rm km\,s^{-1}}$ at $R \sim 300$ kpc.

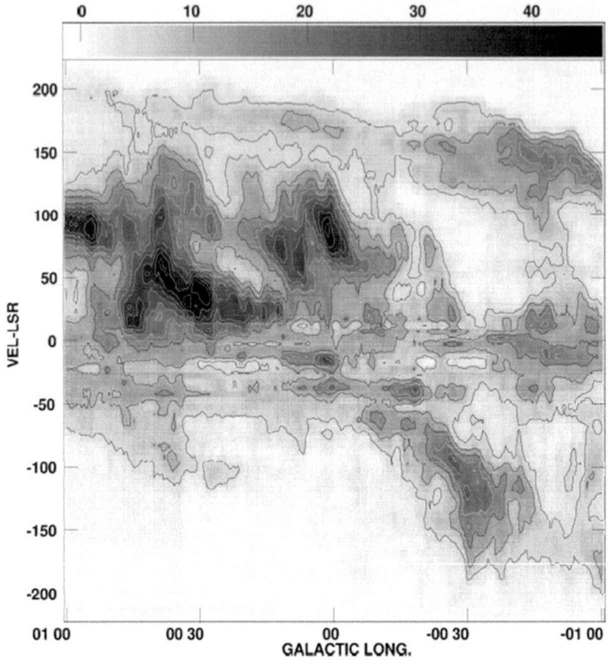

Fig. 4.10 Longitude-velocity diagram across Sgr A of the ^{13}CO($J = 1 - 0$) line emission of the inner 1° region of the galaxy [13]

Fig. 4.11 Inner rotation curve at $R \leq 100$ pc of the galaxy

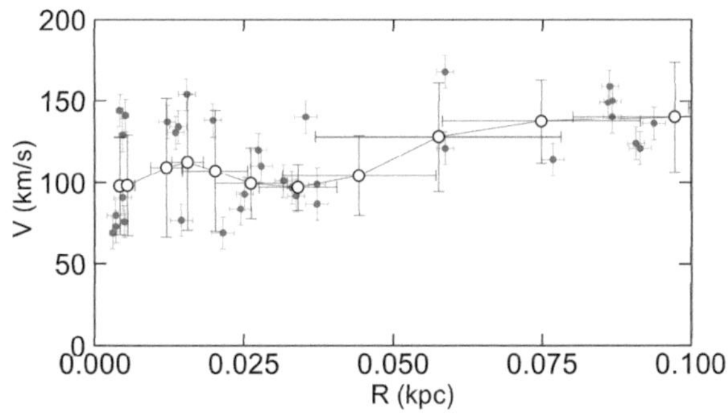

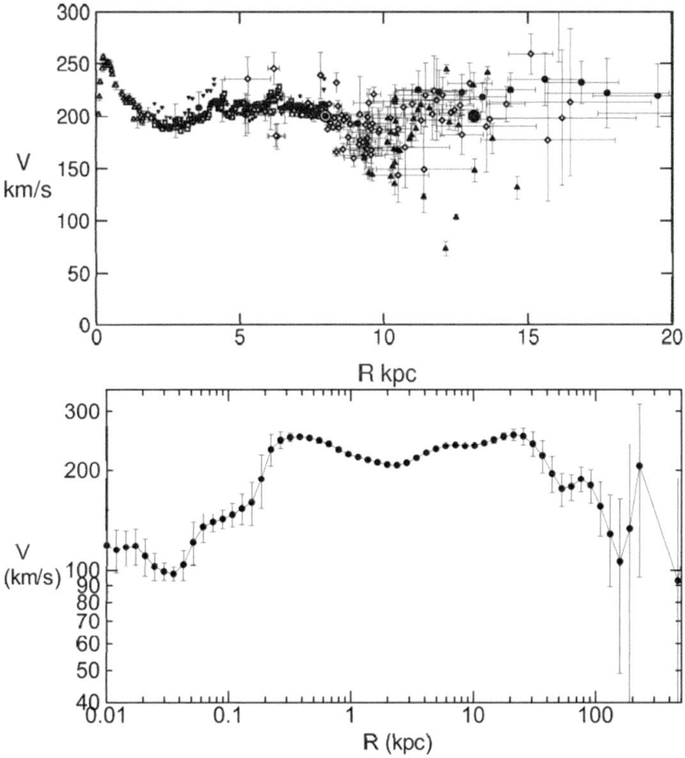

Fig. 4.12 [*Top*] Rotation curve of the galaxy in usual linear plots. [*Bottom*]: Same, but in logarithmic plots in order to see the central region in detail, as well as to view the outermost region for dark halo analysis

4.3 Distribution of ISM

4.3.1 (l, v_r) and (b, v_r) Diagrams

By observing velocity profiles of the HI and/or CO lines at various longitudes, we can derive a diagram in which the intensity (brightness temperature) distribution is mapped in the position-velocity space, namely in the (l, v_r) plane. This is called a position-velocity (PV) diagram. The LV diagram is used for studying the rotation characteristics of the galactic disk as well as to derive the spiral structure. In Figs. 4.13 and 4.14 we show LV diagrams for the HI and CO line emissions as observed along the galactic plane ($b = 0°$).

In a similar way, we can obtain a latitude-velocity diagram by scanning in the direction of the latitude (perpendicular to the galactic plane). A (b, v) diagram is used for obtaining information about the disk thickness and distribution of the gas in the z direction. It is often used for studying the warping of the outer galactic disk.

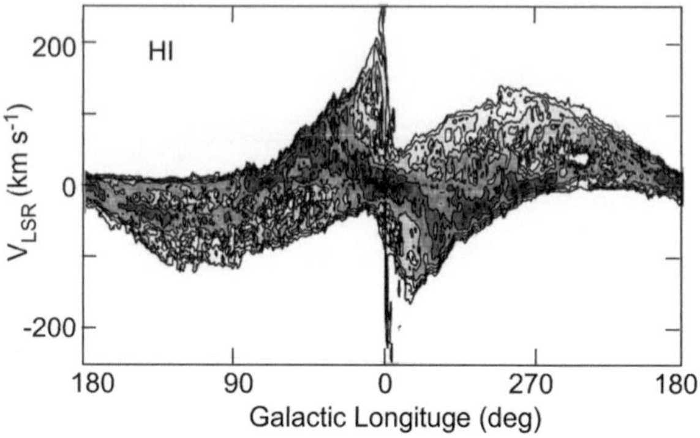

Fig. 4.13 LV diagram for HI gas in the galactic plane [7]

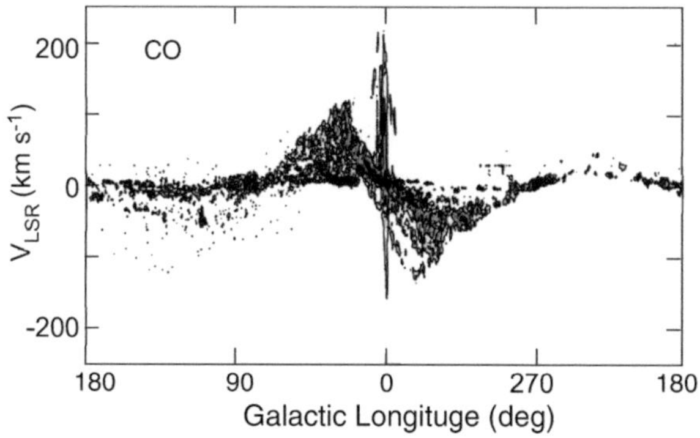

Fig. 4.14 LV diagram for CO (molecular) gas along the galactic plane [3]

Gases with positive velocities represent those inside the solar circle ($R < R_0$), and negative velocities those outside the solar circle ($R > R_0$). A bending of the gas disk is seen in the negative-velocity part (outside the solar circle).

4.3.2 *Velocity-to-Space Transformation Using a Radial-Velocity Diagram*

Given a rotation curve $V(R)$, radial velocity v_r of any object near the galactic plane ($b \sim 0$) is calculated by knowing its (l, r). Inversely, given the radial velocity and galactic longitude (v_r, l) of an object, its distance from the sun and, therefore, its position on the galactic disk, can be calculated. Hence, positions and distribution of objects and gases can be obtained by their line emissions such as recombination lines (HII regions), $\lambda = 21$ cm line emission (HI clouds and diffuse gas), molecular lines such as CO (MCs and GMCs), and maser lines such as SiO (maser stars).

Let the rotation law of the galaxy be $V(R)$; then radial velocity of an object at (R, θ) on (near) the galactic plane is calculated as

$$\begin{aligned} v_{\rm r} &= R\omega \cos(90^\circ - l - \theta) - R_0\omega_0 \sin l \\ &= R\omega(\sin\theta \cos l + \cos\theta \sin l) - R_0\omega_0 \sin l. \end{aligned} \tag{4.29}$$

Because $r \sin l = R \sin l, R \cos\theta = R_0 - r \cos l$, we have

$$v_{\rm r} = R_0(\omega - \omega_0) \sin l = \left\{ \frac{R_0}{R} V(R) - V_0 \right\} \sin l. \tag{4.30}$$

Using Eq. (4.30) we can calculate a radial-velocity diagram of the galaxy, as shown in Fig. 4.15. The radial-velocity diagram is then used to obtain positions (distance r) of an object for which $(l, b; v_{\rm r})$ have been observed:

$$r = f(l, b; v_{\rm r}). \tag{4.31}$$

Positions of objects outside the solar circle are uniquely determined by using this diagram. On the other hand, if the radial velocity is positive at $0 < l < 90^\circ$ and negative at $270 < l < 360^\circ$, the distance of a source is twofold: there are two positions at which the observed $v_{\rm r}$ is satisfied. The discrimination of the near or far side solution is a rather sophisticated problem when obtaining the true distribution of matter. Additional information such as apparent diameters of objects and their structural relation to the surrounding objects is required for the solution.

This procedure is called the velocity-to-space transformation (VST). VST is useful to obtain a face-on map of the density distribution of interstellar mass such as the HI and CO gases in the galaxy. The column density of the HI gas is expressed as

$$N_{\rm HI}\ ({\rm H\ cm^{-2}}) = C \int T_{\rm B}(v) dv\ ({\rm K\ km\ s^{-1}}) \tag{4.32}$$

with $C = 1.82 \times 10^{18}$ being the conversion factor. The local density of the gas is obtained by

$$n_{\rm HI} = \frac{dN_{\rm HI}}{dr} = \frac{dN_{\rm HI}}{dv} \frac{dv}{dr}. \tag{4.33}$$

Using Eq. (4.31) and (4.32), we rewrite this to obtain the VST formula,

$$n_{\rm HI} = C\, T_{\rm B}(v_{\rm r}) \left[\frac{df(l, b; v_{\rm r})}{dv_{\rm r}} \right]^{-1}. \tag{4.34}$$

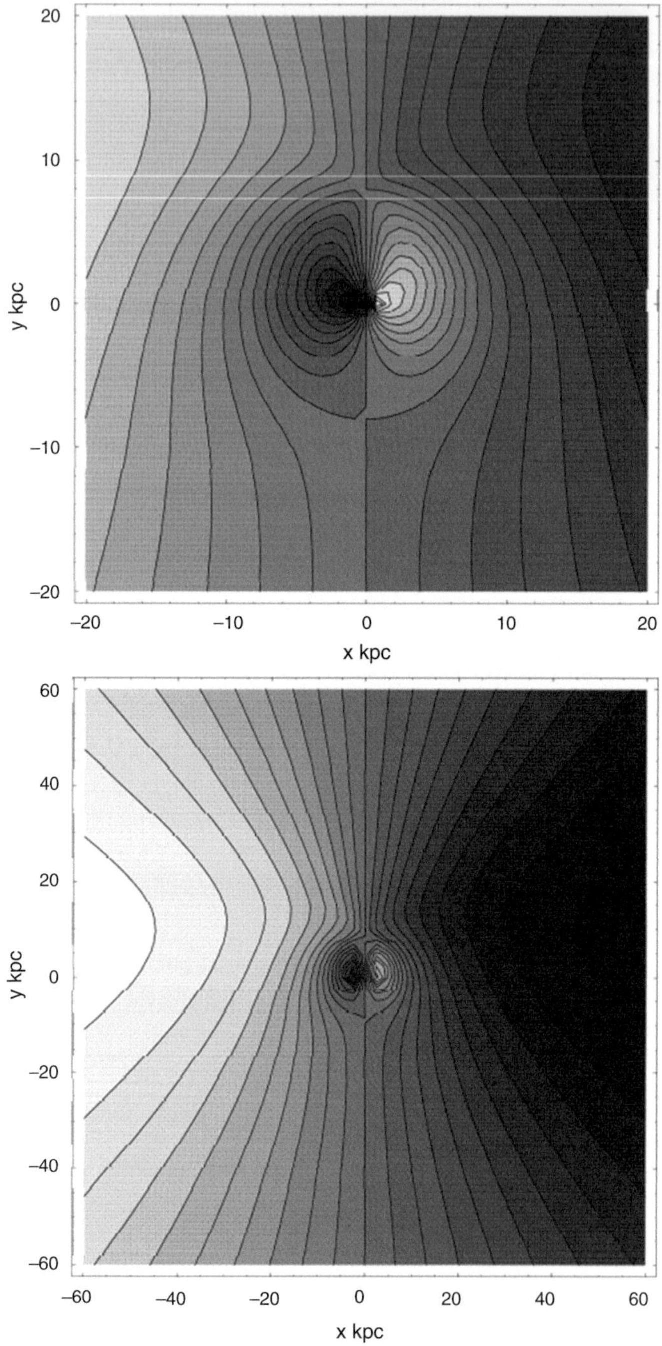

Fig. 4.15 Radial-velocity field of the galaxy at $b \sim 0$

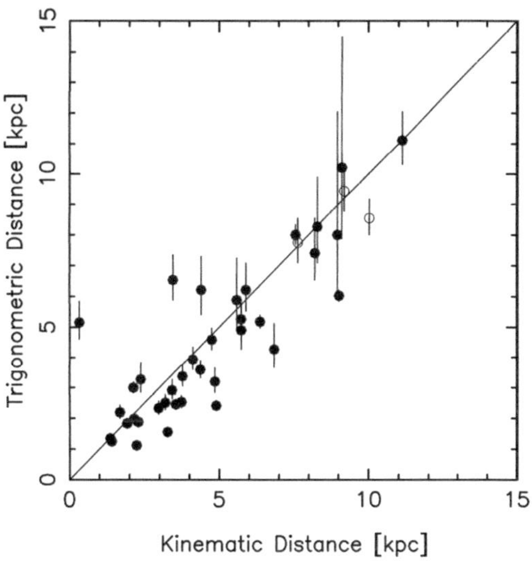

Fig. 4.16 Linear correlation between the kinematic and trigonometric distances determined for galactic maser sources, showing that kinematic distance measurement is reliable

4.3.3 Reliability of Kinematic Distances

The reliability of the kinematic distances used in these analyses can be confirmed by comparing distances of maser sources distributed in the galaxy determined from their radial velocities with their directly determined distances by trigonometric measurement using the VLBI technique. Figure 4.16 shows a comparison of the kinematic and trigonometric distances of galactic objects. The figure shows that both distances are linearly correlated and have good accuracy, confirming that the distances and derived ISM distribution and densities have reasonable accuracy.

4.3.4 Face-On View of the Galaxy

The LV diagram is used for obtaining a face-on distribution of the density of interstellar gas in the galaxy. In Fig. 4.17 we show a face-on view of the galactic disk as seen in the HI line emission obtained by Dutch and Australian astronomers in the 1950s. Several spiral arms can be traced in this diagram. Here, a circular rotation is assumed, and the derived distribution depends on the validity of this assumption.

In the bottom of Fig. 4.17, we show the most recent face-on map of the column density of HI (red) and H_2 (green) gases in the Milky Way. Shown together are a vertical cross-section of the Milky Way in the plane including the galactic center. These maps were obtained from the 3D distribution map of the Milky Way constructed using the HI and CO line observations.

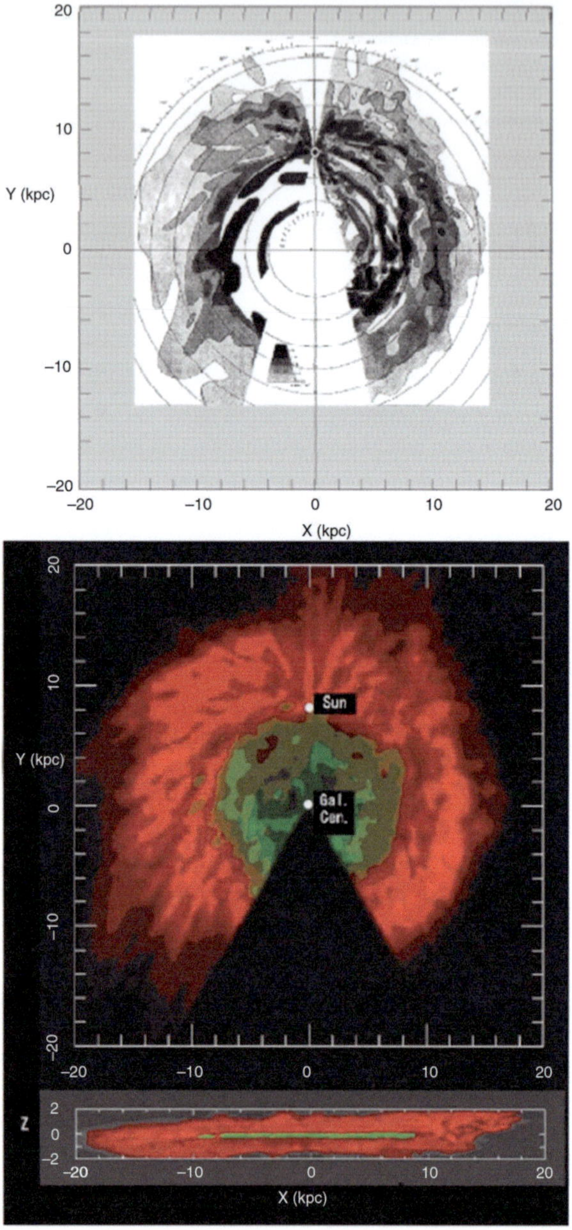

Fig. 4.17 [*Top*] Distribution of HI gas volume density in the galactic plane derived by 21-cm line observations early in the 1950s [14]. [*Bottom*] Projected HI (*red*) and H_2 (*green*) column densities, and cross-section across the galactic center. Molecular gas is concentrated in the inner and near-plane regions [10, 11]

4.3.5 dv/dl Method to Measure Distances to Arms in the GC Direction

In the VST method to map the Milky Way's face-on view, the region near the GC-Sun line ($l \sim 0°$ and $\sim 180°$) is the singular region where the distance equation cannot be applied because of the degeneration of the radial velocity to zero. In this particular region, there is a special method to use the gradient of the LV ridge as the distance indicator. Figure 4.18 illustrates longitude gradients in the LV diagram and corresponding arms and rings, an example of the observed LV diagram in the CO line near the galactic center.

Differentiating Eq. (4.6) by l, we obtain

$$\frac{dv}{dl} = \frac{R_0}{R}(V \cos l - 2Ar \tan p \sin l), \tag{4.35}$$

where p is the pitch angle of the arm, introduced to correct for the tilt of the arm from the rotation circle. However, the second term on the right-hand side can be neglected, and the arms around the GC-Sun line analyzed, where $\sin\ l \ll 1$. Then, the galactocentric distant R of the arm is calculated by

$$R = R_0 \frac{V}{V_0} \left(1 \pm \frac{1}{V_0} \frac{dv}{dl}\right)^{-1}, \tag{4.36}$$

where $+$ and $-$ stand for $l = 0°$ and $180°$, respectively.

If we adopt $V_0 = 200\,\mathrm{km\,s^{-1}}$, this equation can be rewritten as

$$R = R_0 \frac{V}{200\ \mathrm{km\ s^{-1}}} \left(1 \pm 0.286 \frac{dv}{dl°}\right)^{-1} \ (\mathrm{kpc}), \tag{4.37}$$

where $dv/dl°$ is measured in $\mathrm{km\,s^{-1}\,deg^{-1}}$. If we assume a flat rotation curve with $V = V_0$, this equation reduces to a simpler approximation:

$$R \simeq R_0 \left(1 \pm 0.286 \frac{dv}{dl°}\right)^{-1} \ (\mathrm{kpc}). \tag{4.38}$$

In the region close to the GC, where $R \ll R_0$ and the second term in Eq. (4.37) is dominant, this reduces to a further approximation to estimate the radius of the arms and rings:

$$R \simeq R_0 V \left(\frac{dv}{dl}\right)^{-1} = 28 \left(\frac{dv}{dl°}\right)^{-1} \ (\mathrm{kpc}). \tag{4.39}$$

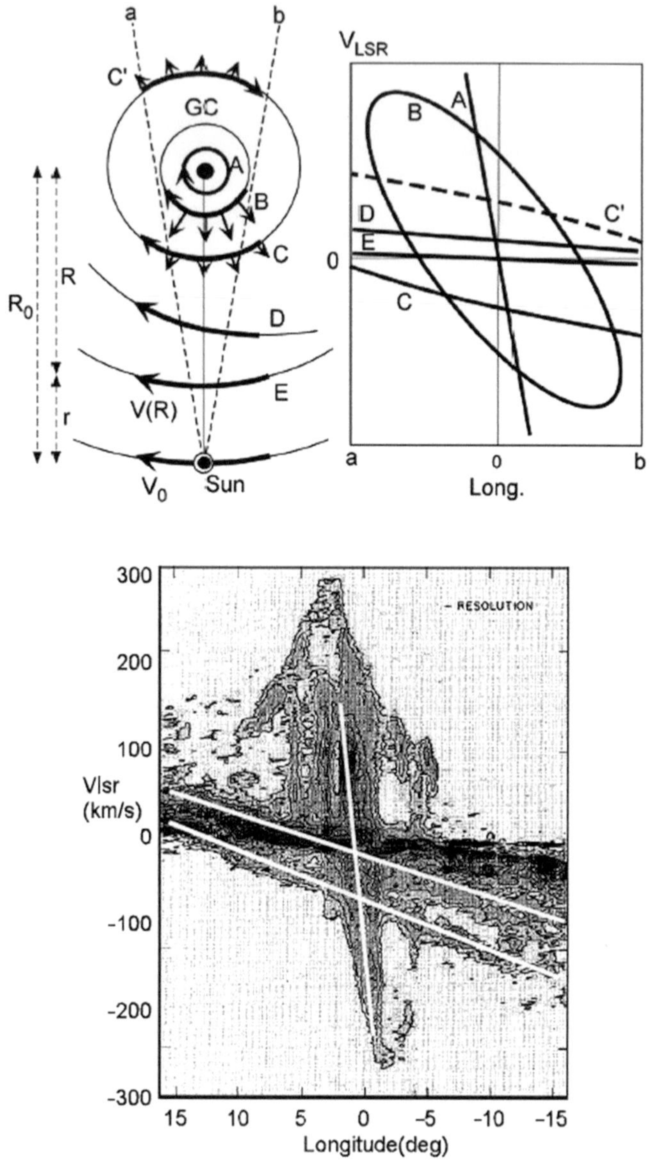

Fig. 4.18 Gradients dv_r/dl of arms and rings traced on the longitude diagram around $l \sim 0°$ are used as a distance indicator [3]

On the other hand, for local arms near the Sun at a distance r, where $R = R_0 + r \sim R_0$, we have

$$r \simeq R_0 \frac{1}{V_0} \left(\frac{dv}{dl} \right) = 2.29 \left(\frac{dv}{dl^\circ} \right) \text{ (kpc).} \tag{4.40}$$

4.3.6 *Spiral Arms in the Milky Way*

Figure 4.19 shows the face-on map of the Milky Way in the HI and CO line emission, displaying superposed spiral arms by tracing the density peaks. The rectangles are positions of spiral arms as determined by the dv/dl method on the Sun-GC line.

The spiral arms traced here may be compared with distributions of the other spiral-arm tracers such as HII regions and SNRs. Figure 4.20 shows an overlay of

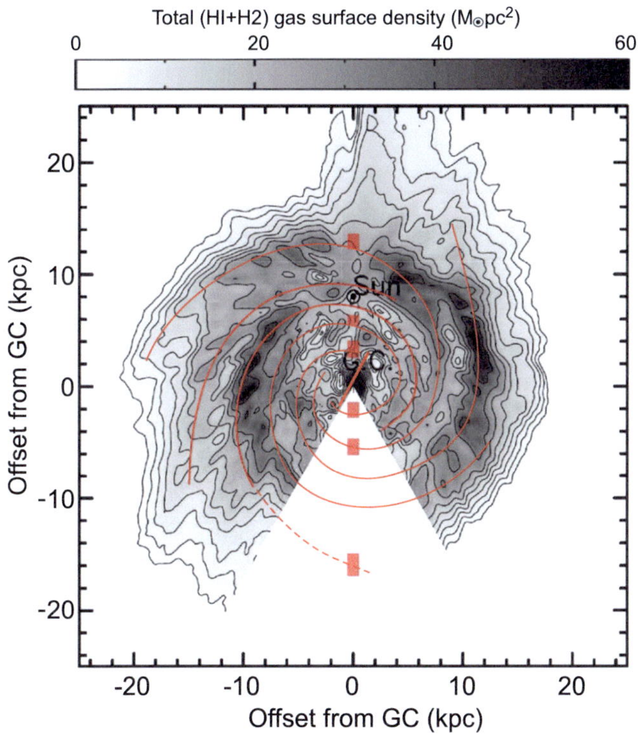

Fig. 4.19 Distribution of the total surface mass density ($HI + H_2$) in the galaxy, as derived by applying the velocity-to-space transformation (VST) [11]. Overlaid are spiral arms traced in the gas distribution. Rectangles indicate positions of spiral arms as determined by the dv/dl method on the GC-Sun line

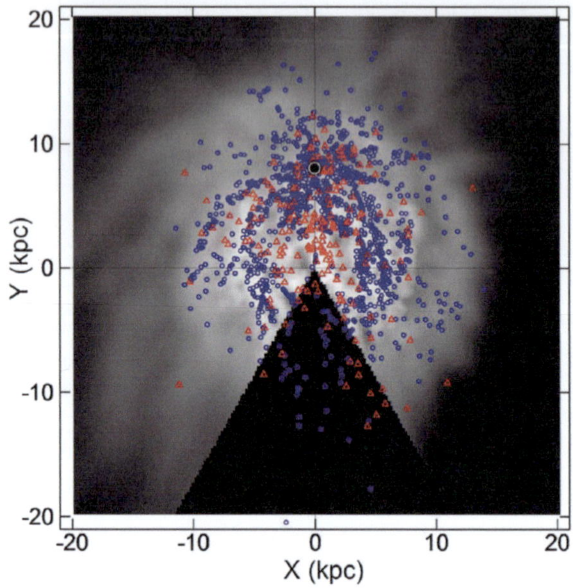

Fig. 4.20 Distribution of HII regions (*blue*: [5]) and SNRs (*red*: [15]) superposed on the face-on map of the total surface gas density (HI + 2H$_2$) of the galaxy [10, 11]

the distributions of HII regions on the face-on surface gas density map, and SNRs with their distances determined from the $\Sigma - D$ relation.

The kinematic distances, and therefore the distribution, of HII regions using their recombination lines are as accurate as those from the HI and CO line emissions. On the other hand, the distances of SNRs are not precise enough to be compared with the HI, CO, or HII kinematic distances because of the rather large scatter in the $\Sigma - D$ relation.

From the traced spiral arms in the figures, we may consider that the Milky Way is an Sb type spiral galaxy. The arms' shape may be approximated by a logarithmic spiral at a pitch angle $p \sim 12°$.

4.4 Density Waves

4.4.1 Gravitational Instability of a Disk

Suppose an infinite plane of stars whose surface mass density and velocity dispersion are σ and $v_{\rm d}$, respectively. Let the disk be perturbed by a density enhancement of radius λ. If the gravity of a mass of the density-enhanced region $M = \pi\lambda^2\sigma$ exceeds the pressure due to the velocity dispersion ρv^2 of a mass element, then the region will contract by self- gravity. Namely, the disk is unstable against gravitational fragmentation if

$$\rho v_{\rm d}^2 < \left(\frac{G\pi\lambda^2\sigma}{\lambda}\right)\rho \tag{4.41}$$

or if

$$\lambda > \lambda_c = \left(\frac{v_d^2}{\pi G \sigma} \right). \tag{4.42}$$

For a galaxy disk for which $v \sim 30\,\mathrm{km\,s^{-1}}$, and $\sigma \sim M_{gal}/(\pi R_0^2) \sim 0.1\mathrm{g\ cm^{-3}}$, the critical wavelength is estimated to be $\lambda_c \sim 0.1$–1 kpc. Therefore, if the galaxy is not rotating, the disk will be unstable against fragmentation into pieces of radius 0.1 to 1 kpc.

However, because the disk is differentially rotating, the effective velocity dispersion is much higher:

$$v^2 \sim v_d^2 + \Delta V_{rot}, \tag{4.43}$$

where

$$\Delta V_{rot} \sim \frac{dV_{rot}}{dR} \lambda \sim (A - B)\lambda = \omega \lambda. \tag{4.44}$$

Even in a case without velocity dispersion ($v_d \sim 0$), we have a larger critical wavelength; the disk is unstable if

$$\lambda > \lambda_c \sim \frac{\pi G \sigma}{(A - B)^2} = \frac{\pi G \sigma}{\omega^2}. \tag{4.45}$$

For a normal disk galaxy the critical wavelength is about $\lambda_c \sim 1$–10 kpc. Therefore, the galaxy disk is unstable for perturbation of a scale length of a few kpc, and the whole galaxy disk suffers from a barlike oval deformation. According to such deformation of the disk shape, and therefore of the disk potential, individual orbits of stars are also deformed to attain "rosette" orbits.

4.4.2 Pattern Speed

There are three kinds of frequencies characterizing the stellar disk: the galactic rotation frequency Ω, the frequency of rotation of the axis of a rosette orbit Ω_p of a star, and the epicyclic frequency κ related to the oscillation of a star in the orbit in the direction of R.

The frequency Ω_p is called also the frequency of pattern rotation, or the pattern (angular) speed. In a normal galactic disk, the pattern speed is approximately given by

$$\Omega_p \simeq \Omega - \frac{1}{2}\kappa. \tag{4.46}$$

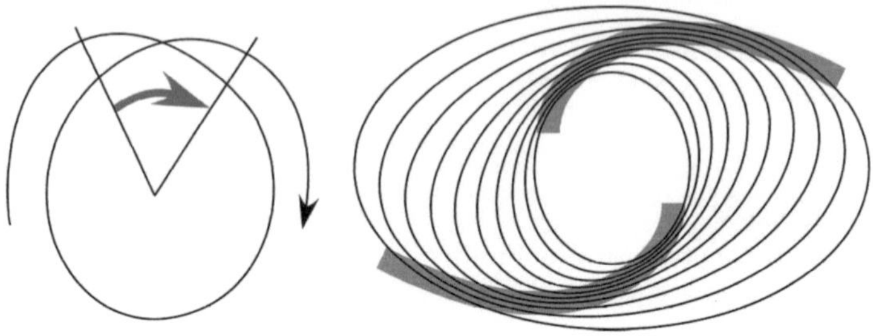

Fig. 4.21 [*Left*] Rosette orbit. [*Right*] Many rosette orbits with shifting major axes in a rotating coordinate make a self-sustaining spiral pattern with their closest-orbit points, producing density wave spiral arms

Therefore, if the disk of stars is perturbed so that all stars begin to rotate on individual rosette orbits, their mutual spatial relation remains about the same for many rotations. Once a "bar" of density-enhanced region is formed, the bar shape is maintained for a long time (Fig. 4.21).

It is a natural consequence that such a distorted bar shape will be affected by the differential rotation, so that the shape is stretched to achieve a spiral pattern. Suppose that the rosette orbits happen to have shifting major axis directions so that orbits of stars become close to each other twice on their orbital motion. Then the orbits and therefore the star distribution attain a self-sustaining spiral pattern along which the potential becomes deeper. This pattern of density enhancement with deeper potential is called the spiral density wave.

Measurement of the pattern speed is not straightforward, because it is the angular velocity of the traveling position of the wave amplitude or the phase velocity of a wave propagating the disk. In fact, stars and interstellar gases are moving through the pattern. Some sophisticated methods have been proposed to measure the speed. A simple method is the offset method, as described in the previous section, and is illustrated in Fig. 4.24.

4.4.3 Resonances

The frequencies $\Omega_{\rm p}$ and κ are related to the galactic parameters as

$$\Omega_{\rm p} = \Omega - \frac{1}{2}\kappa, \tag{4.47}$$

where κ is the epicyclic frequency, related to Oort's constants as

$$\kappa = \sqrt{-A(A-B)}. \tag{4.48}$$

Because of the condition that the pattern speed $\Omega_{\rm p}$ is a slowly varying function of the radius, it can be taken as a constant over the disk. Therefore, once a density wave pattern is formed (either a bar or spiral), the pattern of the potential is maintained for a much longer time than the rotation period of the disk, and the potential rotates rigidly at $\Omega_{\rm p}$.

The pattern velocity near the sun is about $\Omega_{\rm p} R_0 \simeq 100\,{\rm km\,s^{-1}}$. Because the true rotation velocity of stars and gas is $\Omega_0 R_0 = V_0 = 220\,{\rm km\,s^{-1}}$, the density pattern (potential) propagates through the galactic disk at

$$\Delta V_{\rm p} = (\Omega_{\rm p} - \Omega_0) R_0 \sim -120 \text{ km s}^{-1}, \tag{4.49}$$

when it is observed from a rotating frame with the disk. Equivalently, the stars and gas rotate at a relative velocity $-\Delta V = (\Omega_0 - \Omega_{\rm p}) R_0 \sim 120$ km s^{-1} with respect to the density pattern, when observed from a rotating frame with the pattern.

The rotation velocity increases steeply near the galactic center, and is then almost constant over the disk. On the other hand, the pattern velocity $\Omega_{\rm p} R$ increases with R. Therefore a radius appears at which the rotation velocity becomes equal to the pattern velocity, and the guiding center of the rosette orbit rotates with the potential, when

$$\Omega_{\rm p} = \Omega. \tag{4.50}$$

This radius is called the corotation radius, and the corotation resonance occurs there. There occur more resonances, called the Lindblad resonances, when

$$n(\Omega_{\rm p} - \Omega) = \pm\kappa \tag{4.51}$$

with n being integer. At this radius, a star encounters the potential successively and regularly during its epicyclic oscillation. If $\Omega_{\rm p} - \Omega = \kappa/n$, the star overtakes the potential, and this is called an inner Lindblad resonance. If $\Omega_{\rm p} - \Omega = -\kappa/n$, stars are swept by the more rapidly rotating potential, and this is called an outer Lindblad resonance.

4.5 Galactic Shock Waves

4.5.1 *Shocked Stream Line*

Along the density maximum on a density wave pattern, the gravitational potential is deeper compared to the interarm regions. Inasmuch as stars are collisionless, the density enhancement is approximately inversely proportional to the velocity of stars, and the excess depth of the potential due to this density enhancement is considered to be about 10% of the background (unperturbed) potential:

$$\frac{\delta\rho}{\rho} \sim \frac{\delta\Phi}{\Phi} \sim 0.1. \tag{4.52}$$

Velocity excess over the circular rotation is on the order of

$$\frac{\delta V}{V} \sim \frac{1}{2}\frac{\delta \Phi}{\Phi} \sim 0.05, \tag{4.53}$$

or we have $\delta V \sim 10\,\mathrm{km\,s^{-1}}$.

The density wave potential with a spiral pattern of pitch angle p sweeps the interstellar gas at a speed

$$v_{\rm perp} \sim (\Omega - \Omega_{\rm p})R \ \sin\ p. \tag{4.54}$$

In the galaxy the pitch angle is about $p \sim 10°$, and thus $v_{\rm perp} \sim 20\,\mathrm{km\,s^{-1}}$. Because the gas is not collisionless, it happens that gas already swept up and compressed near the center of the potential pattern is encountered by a gas flowing into the arm due to attraction by the potential. The velocity difference between the gas that is about to flow out from the potential and the gas flowing in is a few tens of $\mathrm{km\,s^{-1}}$ in the direction perpendicular to the spiral. This velocity difference is supersonic, and a shock wave occurs at the contact surface of the gases flowing out of and into the potential. This nonlinear response, which causes a shock wave, strongly compresses the interstellar gas and acts to produce stars and bright spiral arms (Figs. 4.22 and 4.23).

Consider a stream line (tube) of gas along an orbit on which the interstellar gas is rotating. If we look at the stream in a rotating frame with the pattern speed $\Omega_{\rm p}$, the gas stream encounters twice the arms per rotation. The encounter of the gas stream occurs at a tilted angle p, and the condition that describes the shock wave is the velocity component of gas perpendicular to the spiral arm

$$v \sim (v_{\rm perp}^2 + \delta V^2)^{1/2} \sim 20\text{–}30\ (\mathrm{km\,s^{-1}}). \tag{4.55}$$

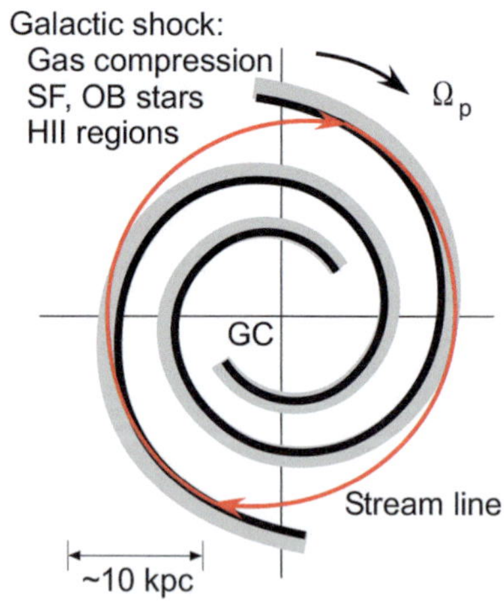

Fig. 4.22 Galactic shock waves

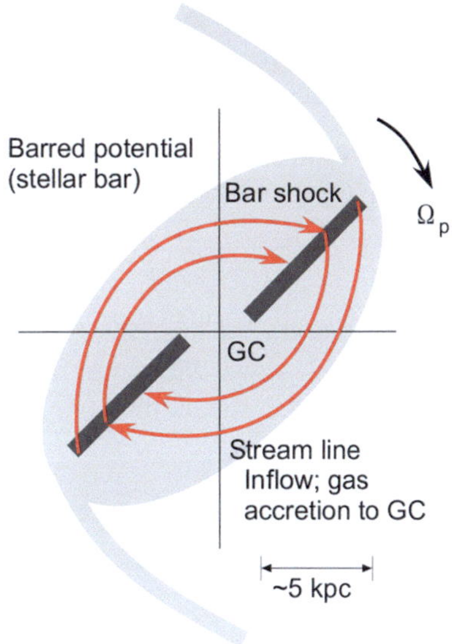

Fig. 4.23 Galactic shock waves in barred potential

Because this velocity is supersonic,

$$M = v/c_s \sim 2\text{–}3, \tag{4.56}$$

a shock wave is generated along the arm in the sense that the compression occurs perpendicular to the arm. On the other hand, the velocity component parallel to the spiral arm has nothing to do with the arm regarding shock wave generation, because it concerns the flow parallel to the arm along which the potential variation is very slow.

4.5.2 *Density and Velocity Jumps and Inflow of Gas*

According to the shock wave, the stream line is bent at the shock. The velocity component perpendicular to the arm drastically changes across the arm in such a sense that the mass flow is conserved:

$$\rho_1 v_1 = \rho_2 v_2, \tag{4.57}$$

where subscripts 1 and 2 denote the status before and after the shock passage. The density jump for an isothermal gas such as the ISM is given by

$$\rho_2/\rho_1 \sim M^2 \sim 10, \tag{4.58}$$

where $M \sim v/c_s \sim 2$–3 is the Mach number, and thus

$$v_2 \ll v_1. \tag{4.59}$$

This results in a loss of kinetic energy as well as in angular momentum of the stream. On the other hand the velocity parallel to the arm does not change:

$$v_{\rm para} \sim (\Omega - \Omega_{\rm p})R \cos p. \tag{4.60}$$

Hence, the stream line becomes oval with its concave part coinciding with the spiral shock wave. Moreover, because of the loss of kinetic energy and angular momentum, the stream line (orbit) does not close, but the gas gradually spirals into the galactic center.

4.5.3 Determination of Pattern Speed

The galactic shock wave has an essential implication for star formation and arm brightening in spiral galaxies. According to the compression of gas along the spiral arm due to the galactic shock wave, the star- formation rate increases by a factor of

$$SFR \propto (\rho_2/\rho_1)^{\alpha} \sim 10^2, \tag{4.61}$$

and $\alpha \sim 1$–2 according to Schmidt's law. The newly born stars are a mixture of massive and less massive stars. Massive stars shine much brighter than less massive stars, but evolve quickly and OB stars shine only for $\sim 10^6$–10^7 y. This means that OB arms can survive only near the dark lane. This is the reason why the spiral arms are observed to be bright along and slightly outside dark lanes in spite of the weak (a few percent) density enhancement of background disk stars (gravitational potential) by the density waves.

There is a time lag between the epoch of cloud formation and that for the formed stars to begin shining. This time lag represents the timescale of star formation, $t_{\rm SF}$, and is related to the pattern speed $\Omega_{\rm p}$, rotation velocity, $\Omega = V_{\rm rot}/R$, and the angular offset Δ between the dark lane and corresponding bright arm (e.g., OB stars and HII regions) along an azimuthal circle, as illustrated in Fig. 4.24. The quantities are related by:

$$\Delta = (\Omega - \Omega{\rm p})\, t_{\rm SF}. \tag{4.62}$$

Here, Ω and Δ can be determined by using the rotation curve and measuring the offset between the dark lane and SF arm along the azimuthal circle, respectively. If measurements are obtained for many positions, we may plot Δ against Ω, as in

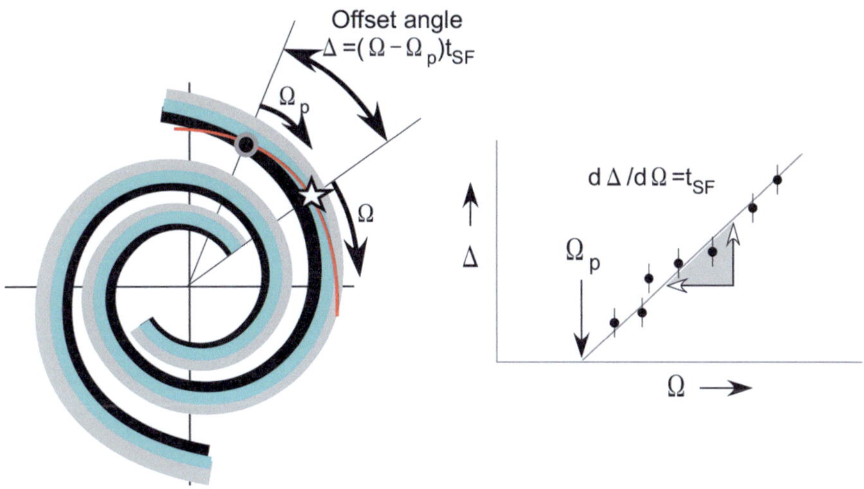

Fig. 4.24 Pattern speed Ω_p and star-formation time t_{SF} are measured simultaneously by plotting offset angles Δ between a dark lane and corresponding SF regions against angular velocity $\Omega = V_{rot}/R$

Fig. 4.24. We may then determine the pattern speed Ω_p as the Ω interception of the linear fit and the gradient,

$$t_{SF} = \frac{d\Delta}{d\Omega}. \tag{4.63}$$

Observations have shown that $\Omega_p \sim 20$–$30\,km\,s^{-1}\,kpc^{-1}$, and $t_{SF} \sim 10^6$ y, which is comparable to the Jeans time in GMC.

4.5.4 Nuclear Fueling by a Bar

The galactic shock wave has also kinematic as well as gas-dynamic implications for the evolution of a gas disk, as well as fueling of the nuclear region by gas. According to the loss of the velocity component perpendicular to the spiral arms the rotation velocity of gas becomes highly noncircular, particularly near the spiral arms. Such a noncircular rotation of gas has been observed as the deformation of iso-velocity contours from those expected for a circular rotation.

The orbit of interstellar gas also shrinks because of this mechanism. If the potential is deep enough, as in such a case for a bar potential, this causes a rapid supply of gas toward the nuclear region of a galaxy. In fact, barred spiral galaxies are shown to experience such gas inflow. This bar-shock contraction of disk gas is considered to be the most efficient fueling mechanism for the nucleus activity and/or starburst in the central region.

4.6 Magnetic and Radio Arms

Magnetic fields are frozen into ISM because of the large magnetic Reynolds number, therefore compression of gas at the galactic shock wave results in compression of the magnetic fields in the direction perpendicular to the arm. By this two-dimensional compression, the strength of the parallel component of the field to the arm is increased by a factor proportional to the gas compression:

$$\frac{B_2}{B_1} \sim \frac{\rho_2}{\rho_1} \sim M^2. \tag{4.64}$$

According to the compression of the magnetic field, the cosmic-ray intensity also increases because of obeying an adiabatic compression law of a relativistic gas:

$$P_{\rm CR} \sim \langle NE \rangle \propto \frac{\rho_2}{\rho_1}^{\gamma} \tag{4.65}$$

with $\gamma = 4/3$ being the adiabatic power index for a relativistic gas. Particle density is proportional to gas density, $N \propto \rho$, therefore

$$\langle E \rangle \propto \rho^{\gamma-1} \sim \rho^{1/3}. \tag{4.66}$$

Thus, the synchrotron emissivity at the arm is increased by a factor

$$\frac{\epsilon_2}{\epsilon_1} \sim \frac{\langle NB^2E^2 \rangle_2}{\langle NB^2E^2 \rangle_1} \sim \left(\frac{\rho_2}{\rho_1}\right) \left(\frac{\rho_2}{\rho_1}\right)^2 \left(\frac{\rho_2}{\rho_1}\right)^{2/3} \sim \left(\frac{\rho_2}{\rho_1}\right)^{3.67} \sim M^{7.3}. \tag{4.67}$$

Therefore, the synchrotron brightness of an arm is by a factor of $M^{7.3} \sim 10^2 \sim 10^3$ times that of the interarm region. In fact, radio continuum observations of spiral galaxies have shown that the radio brightness is much stronger along the spiral arms than in interarm regions, particularly along dark lanes which are places where the shock compression is highest.

4.7 Rotation of Galaxies

4.7.1 Rotation Curves

The rotation curves of galaxies have been obtained by optical (Hα) and HI 21-cm line emission observations along the major axes (Fig. 4.25; for review see [16–18]). It is well known that the HI gas distribution generally shows a depression in the

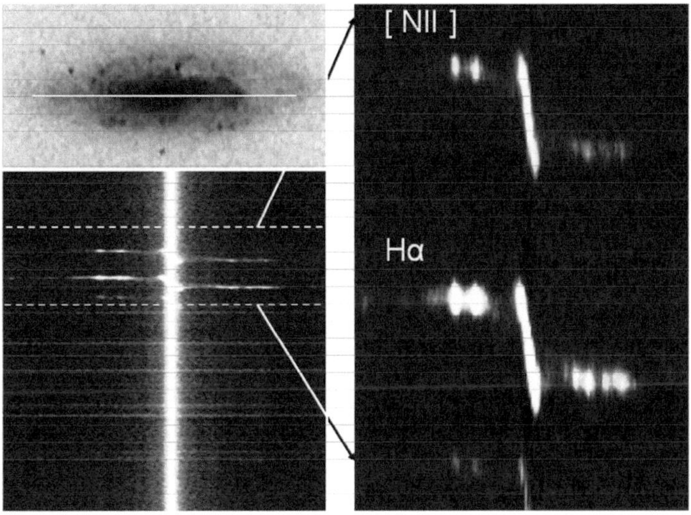

Fig. 4.25 Slit spectra in the Hα and N[II] lines of the spiral galaxy NGC 4257

central few kpc region, which has yielded an apparently solid rotation curve for the central few kpc region. Optical measurements are affected by the contamination of the bright bulge light, which has also increased the uncertainty of the curve near the center. Moreover, because of the dust absorption in the gaseous disk, optical observations cannot be obtained of the central regions of highly tilted galaxies, whereas edge-on galaxies are most suitable for determining the rotation curves without ambiguity of correction for inclination.

On the other hand, the CO-line emission is concentrated in the central region, thus rotation curves of the inner few kpc region can be most accurately obtained by CO position-velocity diagrams. It has been shown that the CO rotation curves of edge-on galaxies do not necessarily coincide with those obtained by HI and/or optical observations. The central rotation by CO is much flatter than that from HI and optical data, or even increases near the center, exhibiting a rapidly rotating compact disk component.

By compiling CO-line PV diagrams along the major axes of nearby late type galaxies that have been observed with large-aperture telescopes and interferometers, the most accurate rotation curves have been derived for nearby galaxies. Particularly, CO-line data from the Nobeyama 45-m telescope with an angular resolution $15''$ have been extensively used. PV diagrams from these data have a linear resolution better than 700 pc (angular resolution $15''$ at 10 Mpc) to 100 pc ($5''$ at 5 Mpc). The inner rotation curves from the CO PV diagrams are combined with outer HI and optical rotation curves from the literature.

4.7.2 Measurement of Rotation Velocity

Figure 4.26 shows an example of the composite PV diagrams for NGC 891. The CO gas is concentrated in the central region, whereas HI is distributed in the outer disks, having a void in the central region. The rotation curve of a galaxy can be derived by using the loci of terminal velocity (V_t) in the PV diagram after correction for the inclination of the galaxy disk. Thereby, the velocity dispersion of the interstellar gas (σ_{ISM}) and the velocity resolution of observations (σ_{obs}) must be corrected by

$$V_{rot} = V_t - (\sigma_{obs}^2 + \sigma_{ISM}^2)^{1/2}. \tag{4.68}$$

The velocity resolution is usually $\sigma_{obs} \sim 10\,\mathrm{km\,s^{-1}}$, and the interstellar velocity dispersion may be taken as $\sigma_{ISM} \sim 7\,\mathrm{km\,s^{-1}}$. Therefore we adopt a correction for the ISM velocity dispersion and velocity resolution as $V_{rot} \simeq V_t - 12\,\mathrm{km\,s^{-1}}$Because the accuracy of measuring V_t is about $10\,\mathrm{km\,s^{-1}}$, the accuracy of determining rotation velocities is not largely dependent on the values of the velocity dispersion and resolution.

The terminal velocity is defined by a velocity at which the intensity becomes equal to

$$I_t = [(0.2 I_{max})^2 + I_{lc}^2]^{1/2} \tag{4.69}$$

on observed PV diagrams. This defines a 20% level of the intensity profile at a fixed position, $I_t \simeq 0.2 \times I_{max}$, if the signal-to-noise ratio is sufficiently high. On the

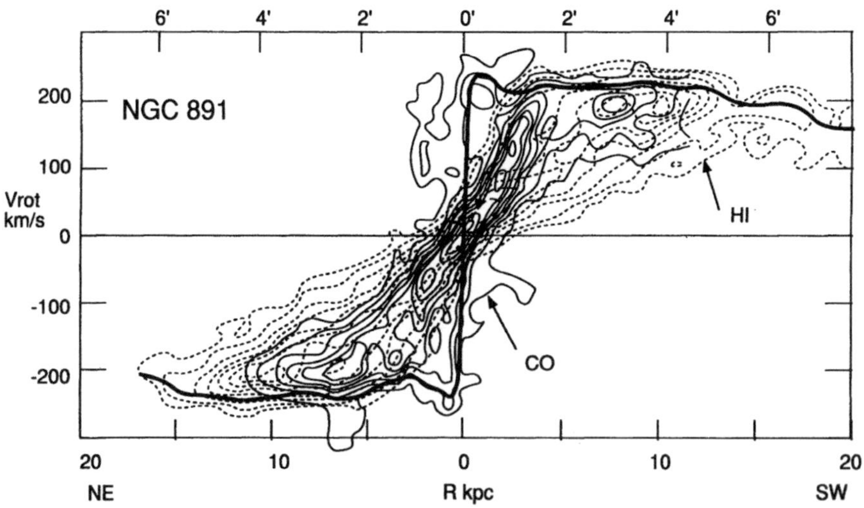

Fig. 4.26 CO + HI composite position-velocity diagram for NGC 891. A fitted rotation curve is superposed by the *thick line*

other hand, if the intensity is not strong enough, the equation gives $I_t \simeq I_{lc}$, which approximately defines the loci along the lowest contour level (usually $\sim 3\times$ rms noise).

In Fig. 4.26 we superpose the thus-obtained rotation curve for NGC 891 as an example. The HI gas indicates the rotation of the outer disk, whereas the CO emission indicates the rotation in the innermost region including the rapidly rotating nuclear disk. The rotation curve as a function of the radius can be obtained by averaging the absolute rotation velocities on both sides of the galaxy nucleus.

4.7.3 Rotation Curves and Galaxy Types

We describe the obtained rotation curves for individual galaxies. Generally, the rotation curves are almost flat even in the very inner region, much flatter than those obtained from HI or optical observations (Fig. 4.27). We describe individual galaxies below. Typical rotation curves observed in nearby spiral galaxies are shown in Fig. 4.28 for Sa, Sb, Sc, and dwarf galaxies.

4.7.4 Radial Mass Distribution for Flat Rotation

Approximately 95% of the mass of the galaxy is possessed by stars, the mass distribution of which can be obtained from kinematic information such as the rotation curve and velocity dispersions. The mass involved within a radius R is given by

$$M(R) \sim \frac{RV(R)^2}{G}. \tag{4.70}$$

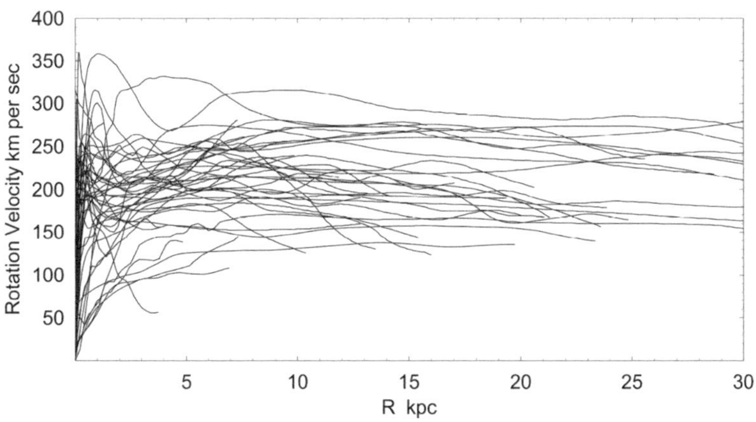

Fig. 4.27 Rotation curves for spiral galaxies

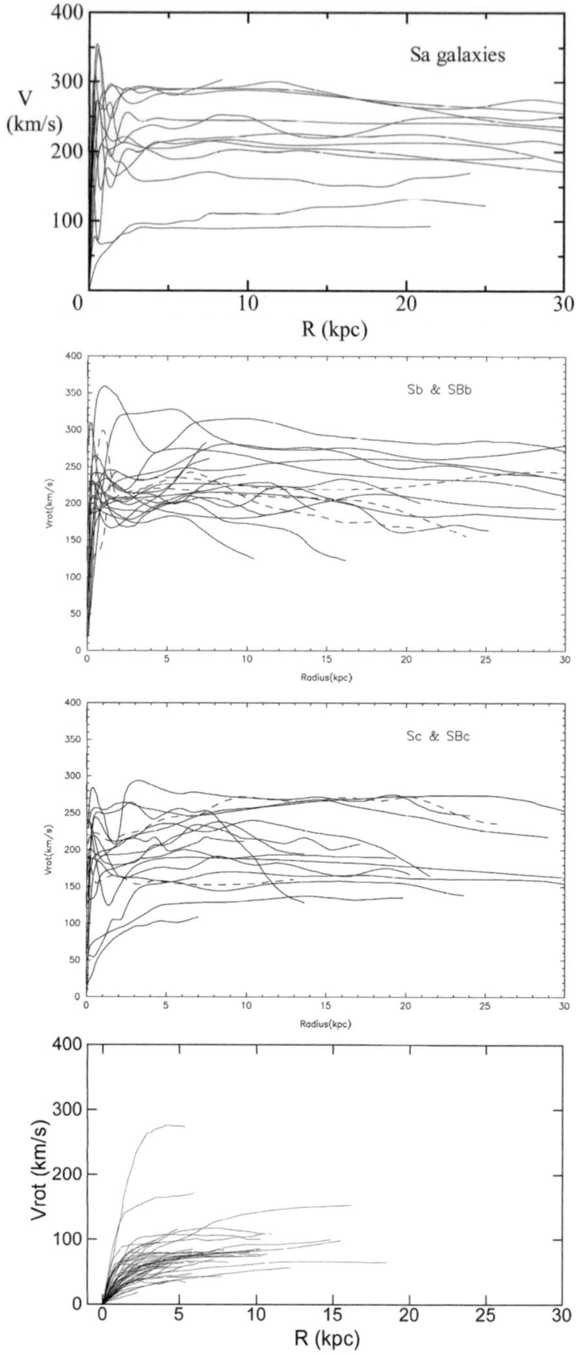

Fig. 4.28 Rotation curves for Sa, Sb, Sc, and dwarf galaxies

If the mass distribution is restricted in the thin surface of an axisymmetric disk, the surface mass density σ is related to the mass $M(R)$ as

$$M(R) \sim \int_0^R \sigma 2\pi r dr, \tag{4.71}$$

or

$$\frac{dM(R)}{dR} = 2\pi R\sigma(R). \tag{4.72}$$

Hence,

$$\sigma \sim \frac{1}{2\pi GR}\left(V(R)^2 + 2RV(R)\frac{dV}{dR}\right) \sim \frac{1}{2\pi GR}V(R)^2, \tag{4.73}$$

where we assumed that $dV/dR \sim 0$. Therefore, the surface density decreases with the radius inversely proportional to R. This may be compared to the case of spherical distribution of mass, in which case the spatial density of mass decreases as $\propto R^{-2}$.

4.7.5 *Mass Distribution Perpendicular to the Disk*

4.7.5.1 Self-Gravitating Stellar Disk

Because the centrifugal force due to rotation balances the gravity by the mass, the local distributions of stars and gas perpendicular to the galactic plane are described by a static equilibrium of the z-directional gravity (perpendicular to the disk) and the velocity dispersion in the z direction v_z.

Generally, the rotation velocity rises steeply within a few hundred pc, indicating the existence of a central compact mass component. Many galaxies (the Milky Way, NGC 891, NGC 3079, NGC 6946) exhibit a sharp maximum at $R \sim$ a few hundred pc, reaching a velocity as high as $\sim$ 200 to 300 km s^{-1}. On the other hand, the maximum velocity corresponding to this component is not high in such galaxies as NGC 253, IC 342, and M51, where the existence of the steep and sharp rise near the nucleus is also evident. One exception is NGC 5907, which shows a gentler (solid-body-like) increase of rotation velocity within $R \sim 3$ kpc.

4.8 Mass Distribution in Galaxies

4.8.1 Approximate Mass of the Galaxy

Because the centrifugal force by the circular orbital rotation of the Sun is in balance with the gravity of the galaxy mass within the solar circle, we may assume that

$$R_0\omega_0^2 \simeq GM_0/R_0^2. \tag{4.74}$$

Or equivalently

$$M_0 = M(R_0) \simeq \frac{R_0 V_0^2}{G}, \tag{4.75}$$

$$or M_0 \simeq \frac{8.5\ \text{kpc}\ (220\ \ \text{km s}^{-1})^2}{G} \sim 1.0 \times 10^{11} M_\odot. \tag{4.76}$$

It is a good approximation to assume a flat rotation for all over the disk: $V(R) \simeq V_0 = 220\,\text{km s}^{-1}$. Then the mass of the galaxy can be expressed as

$$M(R) \simeq 1.0 \times 10^{11} M_\odot \frac{R}{R_0}. \tag{4.77}$$

This implies that the mass of the galaxy increases without limit, which is deeply related to the "missing mass" problem in the outer region.

First we consider a self-gravitating stellar disk. If we denote an equivalent pressure $p = \rho v_z^2$ of the matter concerned, the equilibrium in the z direction can be written as

$$\frac{dp}{dz} = \frac{d(\rho v_z^2)}{dz} = g_z \rho, \tag{4.78}$$

where g_z and ρ are the z-directional gravitational acceleration and the density of the matter (either stars or gases). If we assume that the galactic disk is a flat plane of infinite surface area, the gravity can be approximated as

$$g_z \sim \alpha z, \tag{4.79}$$

and therefore,

$$\frac{d\rho}{dz} \sim -\frac{\alpha z}{v_z^2}\rho. \tag{4.80}$$

This can be solved if the velocity dispersion is constant:

$$\rho = \rho_0 \exp\left(-\frac{z^2}{z_0^2}\right), \tag{4.81}$$

where ρ_0 and z_0 are constants depending on the species. The z_0 is called a z scale height representing the typical thickness of the disk, and is related to the velocity dispersion as

$$z_0 = \frac{v_z}{\sqrt{\alpha}}. \tag{4.82}$$

4.8.1.1 Scale Height of a Gas Disk

Hydrostatic behavior of the interstellar gas is not directly coupled with that of the stellar disk, but is controlled by the gravitational field given by the stellar disk. If we regard the galactic disk as a thin and infinitely extended plane of mass, the z acceleration is written as

$$g_z = 2\pi\Sigma_0, \tag{4.83}$$

where Σ_0 is the surface mass density of the stellar disk. Then we have

$$\frac{d\rho}{dz} = -\frac{2\pi G\Sigma_0}{\mathcal{R}T}\rho, \tag{4.84}$$

where $\mathcal{R}$ is the gas constant and T is the temperature of the gas. Solving this equation, we have

$$\rho = \rho_0 \exp\left(-\frac{z}{z_0}\right), \tag{4.85}$$

where z_0 is the z-directional scale height of the gas disk given by

$$z_0 = \frac{\mathcal{R}T}{2\pi G\Sigma_0}. \tag{4.86}$$

The temperature can be represented by the representative velocity and/or the sound velocity as

$$\mathcal{R}T \sim 1/2v_z^2. \tag{4.87}$$

In Table 4.1 we summarize the typical values of v_z and the disk scale-thickness z_0 for various species.

Table 4.1 Velocity dispersion in the z-direction and scale thickness of the disk

Component	v_z (km s^{-1})	$2z_0$ (pc)
Disk stars	20–30	300
GMC	5	60
OB stars	5	60
HI gas	10	100
Hot gas	$T \sim 10^6$ K	~1 kpc

4.8.2 Axisymmetric Mass Model

The mass distribution $\rho(R, z)$ and the gravitational potential $\Phi(R, z)$ are related by the Poisson equation:

$$\Delta\Phi = -4\pi\rho(R, z), \tag{4.88}$$

Let us recall that the potential for point mass is given by

$$\Phi = \frac{GM}{r} = \frac{GM}{\sqrt{R^2 + z^2}}, \tag{4.89}$$

with $r = \sqrt{R^2 + z^2}$ being the distance from the center. An extended spherical mass is often described by Plummer's law:

$$\Phi = \frac{GM}{\sqrt{r^2 + b^2}} = \frac{GM}{\sqrt{R^2 + z^2 + b^2}}. \tag{4.90}$$

Here, b is a constant representing the scale radius of the spherical mass distribution. The most convenient formula that describes the potential and realistic mass distribution of the galaxy has been obtained by Miyamoto and Nagai [9], and the potential is a modified one from Eq. (4.90) for an axisymmetric spheroid:

$$\Phi = \frac{GM}{\sqrt{R^2 + \left(a + \sqrt{z^2 + b^2}\right)^2}}. \tag{4.91}$$

Here, a is a constant representing the scale radius of the spheroid (disk). The galaxy can be described by three components: the central bulge of spherical distribution, the flattened spheroid corresponding to the disk, and the extended massive halo. The potential of the galaxy is, therefore, a superposition of potentials due to these three mass components:

$$\Phi = \sum_{i=1}^{n} \Phi_i = \sum_{i=1}^{n} \frac{GM_i}{\sqrt{R^2 + a_i + \sqrt{z^2 + b_i^2}}}. \tag{4.92}$$

Table 4.2 Parameters determining the Miyamoto-Nagai potential of the galaxy

Component	M_i ($M_\odot$)	a_i (kpc)	b_i (kpc)
Bulge	2×10^{10}	0.0	0.5
Disk	2.5×10^{11}	7.0	0.5
Massive Halo	$\sim$several$\times 10^{12}$	$\sim$30	$\sim$20

The potential of the galaxy can be fitted by a set of parameters as given in Table 4.2. Using the formula, we obtain the rotation curve and the mass distribution in the galaxy as

$$V_{\rm rot}(R) = \sqrt{\sum_{i=1}^{7} R \frac{\partial \Phi_i}{\partial R}}, \tag{4.93}$$

and

$$\rho(R, z) = \frac{1}{4\pi} \sum_{i=1}^{n} \Delta \Phi_i. \tag{4.94}$$

The third component leads to the flat rotation outside the solar circle.

4.8.3 Decomposition of Rotation Curve

The rotation velocity is calculated by Eq. (4.93). In order to fit the flat rotation at $R \sim$ 10–30 kpc, an extended massive halo has to be introduced, and three-component model ($n = 3$) has been widely used, which takes into account the central bulge, disk, and massive halo. However, after a trial of fitting to the rotation curves of the central few hundred pc region as obtained here, it turned out that the usual three-component model is not sufficient to fit the steep central peak. We have, therefore, introduced a fourth component that represents a more compact nuclear component.

Figure 4.29 shows an example of a calculated rotation curve for a four-component model ($n = 4$), where we assume (1) a nuclear compact mass component, (2) bulge, (3) disk, and (4) a massive halo. Dashed lines indicate rotation curves corresponding to the individual component. The rotation curve of our galaxy, except for the central 10–50 pc, can be fitted by a model with (1) A nuclear mass of $M_1 = 5 \times 10^9 M_\odot$ of a $b_1 = 120$ pc scale radius (2) The bulge of $M_2 = 10^{10} M_\odot$ and $b_2 = 750$ pc radius (3) The disk of $M_3 = 1.6 \times 10^{11} M_\odot$ with radius $a_3 = 6$ kpc and thickness $b_3 = 0.5$ kpc (4) A massive halo of $M_4 = 3 \times 10^{11} M_\odot$ and scale radius of $a_4 = b_4 = 15$ kpc The rotation of NGC 891 can be reproduced by the same model with a similar parameter combination.

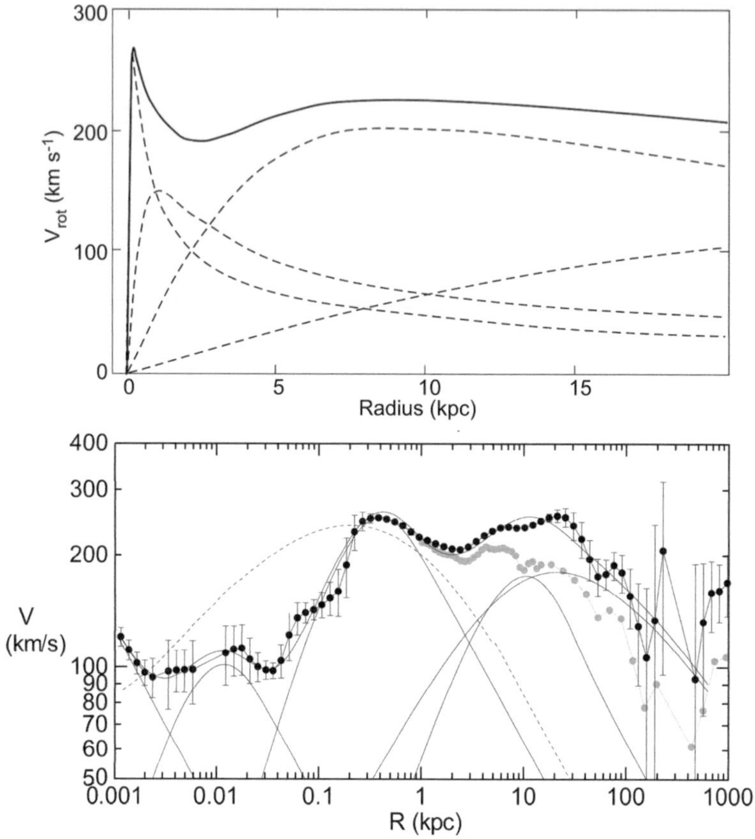

Fig. 4.29 Rotation curve of our galaxy fitted by the Miyamoto-Nagai analytic potential in linear scale (*top*), and by empirical laws for the central BH, two-component exponential bulge, exponential disk, and a dark halo in the logarithmic plot (*bottom*)

Similarly, a rotation curve observed for NGC 6946 is fitted by the one shown in Fig. 4.30, in which a table presenting the parameter combination is inset. The flat valley at $R \sim 1$ to 2 kpc region can be fitted well by introducing the four-component model, which was difficult to reproduce by the three-component model. The rotation of NGC 3079 can be fitted by a similar model. The rotation curves of NGC 253 and IC 342 can be fitted by the same model with a smaller-mass nuclear component, as shown in Fig. 4.30. Rotation curves for the other galaxies can also be reproduced by this model assuming different parameters.

We have thus far called the obtained diagrams the rotation curves, which actually meant observed loci of the highest velocity envelopes in the position-velocity diagrams. However, noncircular motion such as that caused by barred potential and density waves would be superposed on the actual motion of gas, particularly in the central regions. We, therefore, estimate the deviation of the tangential velocity

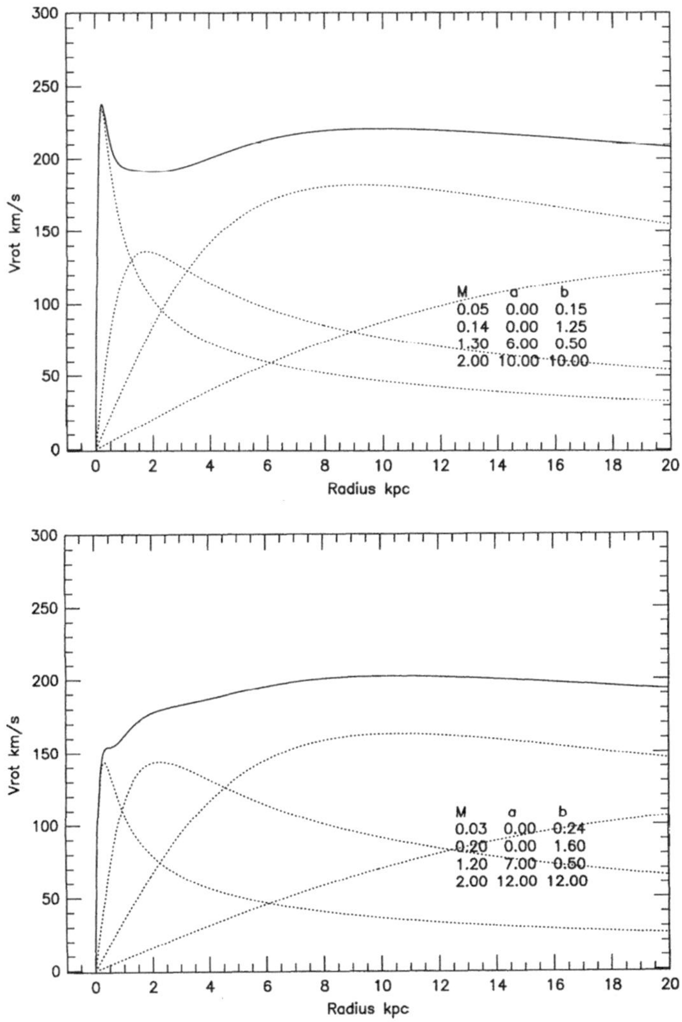

Fig. 4.30 Model rotation curves calculated for the Miyamoto-Nagai potential for spiral galaxies NGC 6946 and IC 342

represented by the observed PV diagrams along the major axis from that of a circular rotation. Suppose that gas clouds are orbiting on elliptical orbits of eccentricity e. Then the orbital velocities of a cloud at the peri- and apogalactic passages are approximated by

$$V_0 = V_{\rm rot}\sqrt{1 \pm e}, \tag{4.95}$$

where $V_{\rm rot}$ is the circular velocity corresponding to the mass distribution. The loci of maximum velocity on the PV diagram will approximate this peri- or apogalactic (maximum or minimum) orbital velocity.

The deviation from circular motion depends on the amplitude and orientation of the arms and bars. That depending on the arms is usually smeared by observing a wide area of the galaxy. Deviation caused by a bar is large and leads to overestimated rotation velocity, when the bar is oriented parallel to the line of sight, whereas it is smaller when the bar is perpendicular. Hence, it is not easy to determine uniquely the mass structure in barred galaxies.

4.8.4 Decomposition by More Empirical Laws

The most impressive feature obtained in the rotation curves is the steeply rising nuclear peak of the rotation curves at $R \sim 100$ to 200 pc, which is generally observed in the central regions. This steep rotation peak can be fitted by a mass model in which a compact nuclear mass component of a 100 to 150 pc radius and a mass of several $10^9 M_\odot$ is assumed. From a fitting of the observed rotation curves this nuclear mass component has turned out to be an additional new component to the well-known central bulge. The rotation curves of galaxies can thus be generally fitted by a model with four mass components: the nuclear compact mass, central bulge, disk, and the massive halo.

The nuclear mass component would have an essential implication for the formation and evolution of the galactic bulge and the central mass condensation of galaxies. A theoretical model has shown that the flat rotation curve and exponential-raw mass distribution in disk galaxies may be a consequence of a viscous protogalactic disk contraction with on-going star formation, where the timescales of viscosity and star formation are of the same order, or of the order of the Jeans time of the disk instability. Their model has also produced a central enhancement of the rotation velocity at $R \sim 0.05R_0$ with R_0 being the scale radius of the disk. This is caused by a raider contraction of the central gas disk compared to the star-formation time because of a stronger shearing-viscosity in the central disk. The model rotation curves could somehow mimic even the central velocity peak of the observed curves such as in the Milky Way, NGC 891, and NGC 6946.

However, the model appears to be still not satisfactory in reproducing in detail the steep central peak of rotation curves at $R <\sim 200$ pc corresponding to the compact nuclear mass component. In order for such a compact mass component to appear, a much stronger raider contraction of protogalactic gas disk would have been necessary. Such a rapid contraction of the gas disk prior to star formation may be possible if we could modify (increase) the viscosity in the central gas disk. Alternatively, we may need to take into account a rapid gas accretion through strong galactic shocks in a central oval (bar) potential during the protogalactic disk contraction.

4.8.5 Dark Matter Halo

The rotation curve of the galaxy has been measured up to the radius $R \sim$ 20–100 kpc, and is known to be almost flat at $V_{\rm rot} \sim 200\,{\rm km\,s^{-1}}$. If we combine this fact with kinematic information about the orbital motion of the satellite galaxies such as the Large and Small Magellanic Clouds, we find that the rotation curve does not decline in a Keplerian law, but maintains a velocity $\sim 100\,{\rm km\,s^{-1}}$ up to $R \sim 300\,{\rm kpc}$. Because the mass involved within radius R is approximately $M(R) \sim RV^2/G$, the fact of a flat or slowly declining rotation indicates that the mass of the galaxy increases with the radius without limit.

The mass-to-light ratio in the inner region of a galaxy is roughly constant for normal galaxies up to a certain radius. On the other hand, total luminosity of a whole galaxy is, of course, finite: the surface brightness of a galaxy decreases with the radius as the surface density of mass decreases, and it reaches a very low level at a certain radius $R_{\rm H}$ beyond which no luminous matter is observed. As the radius increases, the luminosity attains its maximum value at a certain radius beyond which the luminosity does not increase. Therefore, it happens that the mass-to-light ratio $M(R)/L(R)$ begins to increase proportionally to R beyond this radius, $M(R)/L(R) \sim M(R)/L(R_{\rm H}) \propto R$.

There exists a huge amount of mass in the space surrounding the galaxy, even though there is no light at all. The huge volume of the halo is filled with "dark matter," and the space is called the "dark halo." In spite of efforts to look for any emission (from X-rays to radio), the mass has been not detected, and is now a major subject for direct detection as a fundamental particle in laboratory experiments. The dark matter problem has been one of the most intriguing mysteries in astrophysics with an expected challenge in both theoretical and experimental fundamental physics.

Figure 4.31 shows edge-on views of the surface mass density (SMD) or the integrated mass along the lines of sight, of a typical galaxy, as calculated for observed rotation curves. The total mass of the galaxy is dominated by the extended dark matter halo.

4.8.6 Mass Models for Rotation Curve Decompositions

Here we present some useful functions often used to represent the galactic rotation curves. Models for the bulge and disk have been proposed based on optical surface brightness photometry of many galaxies. Dark halo models have been found from the flat rotation in the halo as well as from numerical simulations of galaxy evolution in the expanding universe.

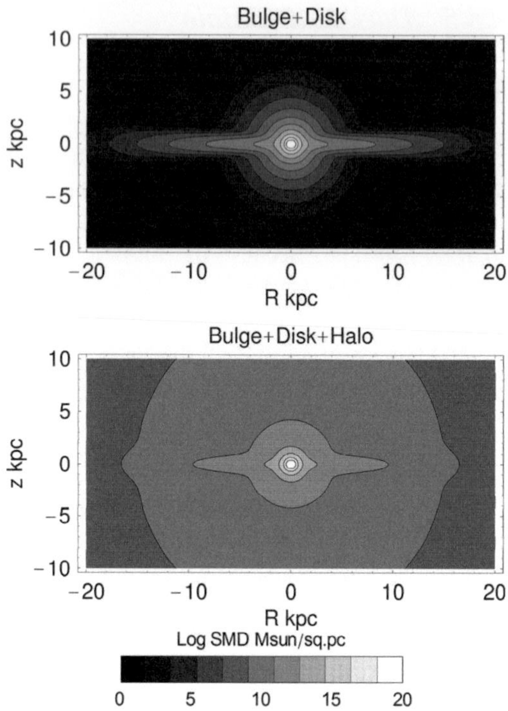

Fig. 4.31 Edge-on view of the SMD (surface mass density) of the bulge and disk (*top*), and bulge, disk, and dark halo (*bottom*). The total mass is dominated by the dark halo mass

4.8.6.1 de Vaucouleurs Bulge

A commonly used profile to represent the central bulge is the de Vaucouleurs law (Fig. 4.32) for surface-brightness distribution, which is expressed by the projected radius R from the center by

$$\log\beta = -\gamma(\alpha^{1/4} - 1), \tag{4.96}$$

where $\gamma = 3.3308$, $\beta = B_{\rm b}(R)/B_{\rm be}$, $\alpha = R/R_{\rm b}$, and $B_{\rm b}(R)$ is the surface-brightness normalized by the value at radius $R_{\rm b}$, $B_{\rm be}$.

The same function is adopted for the surface mass density as

$$\Sigma_{\rm b}(R) = \lambda_{\rm b} B_{\rm b}(R) = \Sigma_{\rm be}\exp\left[-\kappa\left(\left(\frac{R}{R_{\rm b}}\right)^{1/4} - 1\right)\right], \tag{4.97}$$

where $\Sigma_{\rm bc} = 2142.0\Sigma_{\rm be}$ for $\kappa = \gamma\ln 10 = 7.6695$, and $\lambda_{\rm b}$ is the M/L ratio assumed to be constant.

Equations (4.96) and (4.97) show that the central SMD at $R = 0$ attains a finite value, and decreases steeply outward near the center. However, the decreasing rate becomes milder at large radii, forming an extended outskirt.

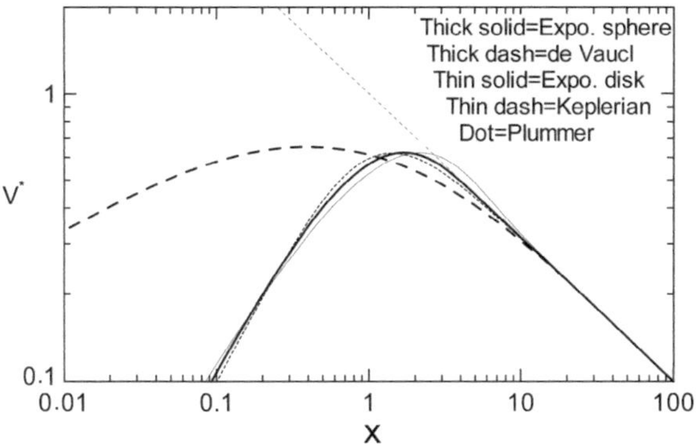

Fig. 4.32 Comparison of normalized rotation curves for the exponential spheroid, de Vaucouleurs spheroid, and other typical models, for a fixed total mass. The exponential spheroid model is almost identical to that for the Plummer law model. The central rise of the de Vaucouleurs curve is proportional to $\propto r^{1/2}$, whereas the other models show a central velocity rise as $\propto r$

The cylindrical mass inside R is calculated as

$$M_{\mathrm{b:cyl}}(R) = 2\pi \int_0^R x\Sigma_{\mathrm{b}}(x)dx. \tag{4.98}$$

Total mass of the bulge given by

$$M_{\mathrm{bt}} = 2\pi \int_0^\infty R\Sigma_{\mathrm{b}}(R)dR = \eta R_{\mathrm{b}}^2 \Sigma_{\mathrm{be}} \tag{4.99}$$

with $\eta = 22.665$. Half of the total projected (cylindrical) mass is equal to that inside a cylinder of radius R_{b}. The volume mass density $\rho(r)$ at radius r for a spherical bulge is now given by

$$\rho(r) = \frac{1}{\pi} \int_r^\infty \frac{d\Sigma_b(x)}{dx} \frac{1}{\sqrt{x^2 - r^2}} dx. \tag{4.100}$$

The rotation velocity is given by the Kepler velocity of the spherical mass inside R as

$$V_{\mathrm{b}}(R) = \sqrt{\frac{GM_{\mathrm{b:sph}}(R)}{R}}, \tag{4.101}$$

where

$$M_{\mathrm{b:sph}}(R) = 4\pi \int_0^R r^2 \rho(r)dr. \tag{4.102}$$

It must be remembered that the spherical mass $M_{\mathrm{b:sph}}(R)$ is smaller than the cylindrical mass $M_{\mathrm{b:cyl}}(R)$ given by equation (4.98). At large radii, the velocity approximately decreases by the Keplerian law. Figure 4.32 shows the variation of circular velocity for a de Vaucouleurs bulge.

4.8.6.2 Exponential Disk

Optical observations of galactic disks revealed that the surface brightness is well represented by an exponential function. If the M/L ratio in the disk does not vary strongly, the mass distribution may be approximated by the same function, and may be written as

$$\Sigma_{\mathrm{d}}(R) = \Sigma_{dc}\exp(-R/R_{\mathrm{d}}). \tag{4.103}$$

Here, Σ_{dc} is the central value, R_d is the scale radius, and the total mass of the disk is given by $M_{\mathrm{disk}} = 2\pi\Sigma_{dc}R_{\mathrm{d}}^2$. The rotation curve for a thin exponential disk is expressed by

$$\begin{aligned} V_{\mathrm{d(R)}} &= \sqrt{R\frac{\partial\Phi}{\partial R}} \\ &= \sqrt{4\pi G\Sigma_0 R_{\mathrm{d}} y^2[I_0(y)K_0(y) - I_1(y)K_1(y)]}, \end{aligned} \tag{4.104}$$

where $y = R/(2R_{\mathrm{d}})$, and I_i and K_i are the modified Bessel functions.

More generally, in a case where the surface density does not obey the exponential law, the gravitational force $f(R)$ can be calculated by integrating the x directional force caused by mass element $\Sigma_{\mathrm{d}}'(x)dxdy$ in the Cartesian coordinates (x, y):

$$f(R) = G\int_{-\infty}^{\infty}\int_{-\infty}^{\infty}\frac{\Sigma_d(x)(R-x)}{s^3}dxdy, \tag{4.105}$$

where $s = \sqrt{(R-x)^2 + y^2}$. The rotation velocity is then given by

$$V_d(R) = \sqrt{fR}. \tag{4.106}$$

This formula is used for any thin disk with an arbitrary SMD distribution $\Sigma(x, y)$ in cases where the disk is superposed by spiral arms, rings, clumps, or a bar.

4.8.6.3 Isothermal Halo

Outermost rotation curves are observed to be flat in most galaxies, and semi-isothermal spherical distribution is often used to fit the curves, where the mass density is expressed by

$$\rho_{\text{iso}}(r) = \frac{\rho^0_{\text{iso}}}{1 + (r/h)^2}, \tag{4.107}$$

where ρ_{hc} and $h = r_{\text{h}}$ are the central mass density and scale radius, respectively. The circular velocity is given by

$$V_h(R) = V_\infty \sqrt{1 - \left(\frac{h}{R}\right) \tan^{-1}\left(\frac{R}{h}\right)}, \tag{4.108}$$

which approaches a constant rotation velocity V_∞ at large distances. Note that $r = R$ in this spherical case for the halo.

At a small radius, $R \ll h$, the density becomes nearly constant equal to ρ^0_{iso} and the enclosed mass increases steeply as $M(R) \propto R^3$. At large radii, the density decreases with $\rho_{\text{iso}} \propto R^{-2}$. The enclosed mass increases almost linearly with the radius as $M(R) \propto R$.

4.8.6.4 Navarro-Frenk-White (NFW) Model

The most popular model for the dark halo is the NFW model empirically obtained from numerical simulations in the cold-dark matter scenario of galaxy formation [12]. In the model the mass density distribution is expressed by

$$\rho_{\text{NFW}}(r) = \frac{\rho^0_{\text{NFW}}}{(r/h)/[1 + (r/h)^2]}. \tag{4.109}$$

The circular velocity is calculated by

$$V_h(R) = \sqrt{\frac{GM_{\text{h}}(R)}{R}}, \tag{4.110}$$

where M_h is the enclosed mass within the scale radius h.

In the central region at $R \ll h$ the density behaves as $\rho_{\text{NFW}} \propto 1/R$, infinitely increasing toward the center, making a central cusp there, and the enclosed mass behaves as $M(R) \propto R^2$.

At large radii, it behaves as $\rho_{\text{NFW}} \propto R^{-3}$, yielding a logarithmic increase of mass, $M(R) \propto \ln R$.

4.8.7 *Direct Calculation of Mass Distribution from Rotation Curves*

The mass distribution in a galaxy can also be obtained by directly calculating it using the rotation curve [2]. In this case, no functional form is necessary to represent the mass a priori, as above. However, an assumption has to be made as to whether the galaxy's shape is spherical or a flat disk, considering that the true mass profile may lie between these two extreme cases. Figures 4.33 and 4.34 shows examples of the calculated mass density profiles for galaxies using their observed rotation curves. In the following subsections we describe the methods to calculate the surface mass density under these assumptions.

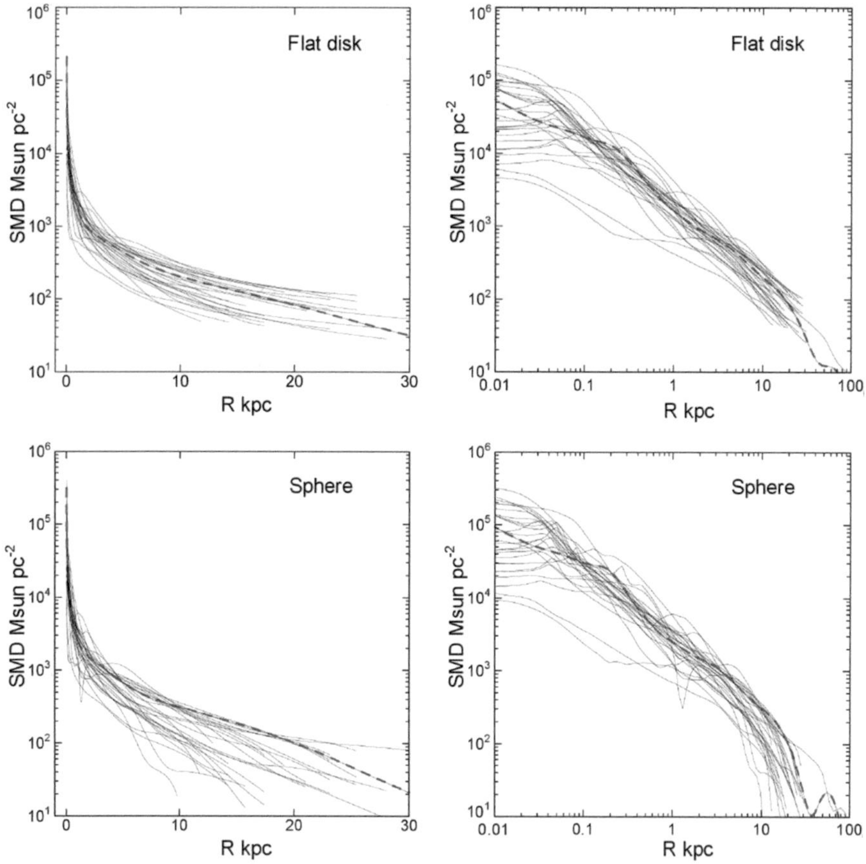

Fig. 4.33 [*Upper panels*] Direct SMD in spiral galaxies with end radii of RC greater than 15 kpc calculated under flat disk assumption, and the same in logarithmic radius [17]. *Red dashed lines* indicate the Milky Way. [*Lower panels*] Same but under sphere assumption

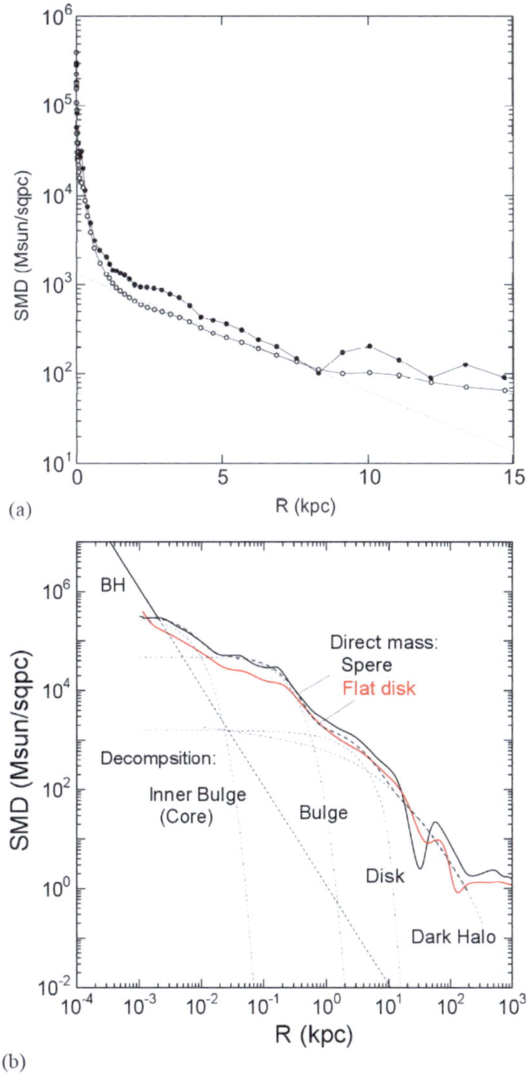

Fig. 4.34 (**a**) Radial distribution of SMD in the Milky Way calculated directly using the rotation curve assuming spherical (*black dots*) and flat-disk (*open circles*) density distributions. (**b**) Same but in log-log plots (*black* for sphere; *red* for disk). *Dashed curves* are the results by decomposition method of rotation curve into black hole, bulge, disk, and dark halo

4.8.7.1 Sphere

On the assumption of spherical distribution, the mass inside radius r is given by

$$M(r) = \frac{rV(r)^2}{G}. \tag{4.111}$$

Then the surface mass density $\Sigma_S(R)$ at R is calculated by

$$\Sigma_S(R) = 2\int_0^\infty \rho(r)dz. \tag{4.112}$$

Remembering

$$\rho(r) = \frac{1}{4\pi r^2}\frac{dM(r)}{dr}, \tag{4.113}$$

the above expression can be rewritten as

$$\Sigma_S(R) = \frac{1}{2\pi}\int_R^\infty \frac{1}{r\sqrt{r^2 - R^2}}\frac{dM(r)}{dr}dr. \tag{4.114}$$

The spheroidal component in the central region is well represented by this method, whereas the disk mass density is underestimated, particularly near the edge of the data points at the maximum radius of observations.

4.8.7.2 Flat-Disk

The SMD of a flat thin disk, $\Sigma_D(R)$, is calculated by solving the Poisson equation assuming a disk of negligible thickness, and is given by

$$\Sigma_D(R) = \frac{1}{\pi^2 G}\times$$

$$\left[\frac{1}{R}\int_0^R \left(\frac{dV^2}{dr}\right)_x K\left(\frac{x}{R}\right)dx + \int_R^\infty \left(\frac{dV^2}{dr}\right)_x K\left(\frac{R}{x}\right)\frac{dx}{x}\right]. \tag{4.115}$$

Here, K is the complete elliptic integral, which becomes very large when $x \simeq R$.

4.8.8 Velocity Field and Galaxy Warp

Spectral HI data provide us with kinematic information of disks of spiral galaxies. The variation of radial velocity v as a function of the distance from the center along the major axes gives almost directly the rotation curve $V(R)$:

$$V(R) = v/\sin\ i, \tag{4.116}$$

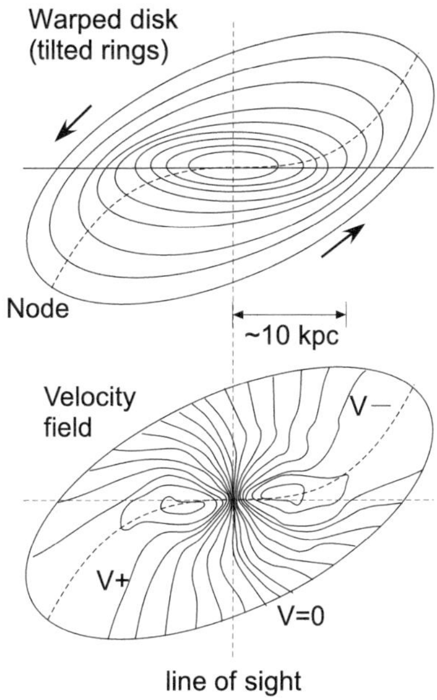

Fig. 4.35 Warping of galactic disk, represented by a tilted ring model, and velocity field superposed by perturbed velocities such as caused by spiral arms

where i is the inclination angle of the galaxy ($i = 0°$ for a face-on galaxy; $i = 90°$ for edge-on). The radial velocity distribution as plotted in the form of iso-velocity contours is called the velocity field. A velocity field gives information about not only the rotation but also such noncircular motions caused by spiral arms (density waves and galactic shock waves) and by warping of the disk.

Figure 4.35 shows an example of a velocity field and corresponding illustration of tilted rings in a warped disk galaxy. The nodal line as defined by loci of maximum (recession side of the disk rotation) and minimum (approaching side) velocity is often bending. This is caused mainly by large-scale warping of the disk, particularly in the outermost regions. Also wavy variation of the iso-contours is recognized in almost every galaxy, and is caused by noncircular motions superposed on the circular rotation because of density wave and galactic shocks.

4.9 Evidence for Dark Matter in Galaxies

4.9.1 Flat Rotation Curves

Flat rotation similar to that of our galaxy has been observed for almost all galaxies for which rotation curves were measured in the HI line and optical line emissions such as the Hα line. In these galaxies, the mass increases with the radius almost infinitely, and the luminosity is finite, and again the missing-mass problem has been a challenging theoretical subject such as the cold-matter scenario.

4.9.2 Binary Galaxies

The mass of double galaxies and a galaxy with a companion (satellite) can be estimated by relative velocities of the two galaxies and their separation by assuming the inclination of the orbit. The orbital velocity of a galaxy around the other is expressed as

$$V_{\rm rot} \sim \frac{1}{\sin\ i}(v_1 - v_2), \tag{4.117}$$

where i is the inclination of the orbital axis from the line of sight, and v_1 and v_2 are the radial velocities of the two galaxies. The reduced mass of the two galaxies is related to the orbital velocity as

$$V_{\rm rot}^2 \sim G\left(\frac{m_1 m_2}{m_1 + m_2}\right)\frac{1}{R}, \tag{4.118}$$

where m_1 and m_2 are the masses of the two galaxies, and R their separation.

4.9.3 Virial Mass of a Cluster of Galaxies

Masses of groups of galaxies and clusters of galaxies are estimated by the Virial mass from their size and velocity dispersion. Clusters of galaxies contain thousands of galaxies that are gravitationally bound to the center of mass of the cluster. The Virial mass (dynamical mass) of a cluster is given by

$$M_{\rm c} \sim \frac{R_{\rm c} v_\sigma^2}{G} \tag{4.119}$$

where R and v_σ are the radius of the cluster and velocity dispersion of member galaxies. Observations have shown that $v_\sigma \sim 2000\,{\rm km\,s^{-1}}$, and $R_{\rm c} \sim 1\,{\rm Mpc}$ for a normal-size cluster, and therefore, $M_{\rm c} \sim 8 \times 10^{14} M_\odot \sim 10^4 M_{\rm Galaxy}$. However, the total mass of individual galaxies cannot explain this huge amount of dynamical mass.

4.9.4 Dark Matter

The majority of the mass contained by a cluster is not visible as galaxies. Hence, clusters of galaxies also contain invisible mass that amounts to about 10 times the mass of all member galaxies; a cluster of galaxies is filled with missing mass.

Therefore, we may consider that the "luminous mass" of galaxies that are observable by classic astronomical means is much smaller than the dynamic mass of galaxies. We are "seeing" only a very small fraction ($\sim$1–10%) of the real mass which may be either in a form such as invisible small (dark) stars or in some exotic form often known as cold-matter.

4.10 Tully-Fisher Relation

4.10.1 HI Tully-Fisher Relation

Rotation velocity and size of a galaxy are directly related to the galaxy mass, and therefore, to the galaxy luminosity [19].

$$L \propto M = \frac{RV_{\rm rot}^2}{G}. \tag{4.120}$$

If the galaxy luminosity is related to the surface area of the disk as

$$L \propto R^2, \tag{4.121}$$

or

$$R \propto M^{1/2}, \tag{4.122}$$

then, Eq. (4.120) reduces to

$$L \propto M \propto V_{\rm rot}^4. \tag{4.123}$$

Or, using absolute magnitude $M_{\rm B}$ in the B band, for example, we obtain

$$M_B \sim -10 \log V_{\rm rot} + {\rm const.} \tag{4.124}$$

Indeed, a good relation has been found between the total velocity width of the HI line emission and the absolute magnitude. In reality, however, an empirical relation has been obtained for nearby galaxies with calibrated distances:

$$M_{\rm B} = -7.58(\log W_i - 2.50) - 19.55 \tag{4.125}$$

where $W_i = 2V_{\rm rot} = W_{\rm obs}/\sin\ i$ is the line width after correction for the inclination of the galaxy. This relation is called the "Tully-Fisher relation (TFR)."

The Tully-Fisher relation is one of the most powerful tools to estimate distances to galaxies using total line widths of the 21-cm HI line emission. However, distances to galaxies reached thus far by HI observations are limited to around 100 Mpc, or recession velocities of 10,000 to 15,000 km s^{-1} even with use of the largest telescope.

4.10.2 CO Versus HI Line Profiles

Calibration of the CO TFR can be obtained by confirming that the CO line profiles have been found to coincide with the HI profiles in nearby galaxies such as NGC 891 and NGC 4565. Generally, rotation curves are almost flat beyond 1 kpc from the galactic center, attaining a broad maximum at around the galactocentric distance of $R \sim$ 3–5 kpc. The rotation and/or dispersion is higher in the central few hundred pc, which contributes to the high-velocity wing in the total profile. The radial distribution of CO gas has two major density maxima: one near the center and the other at around $R \sim$ 3.4 and 4.5 kpc for NGC 891 and NGC 4565, respectively. When the line emission is integrated over an entire galaxy, the contribution from the outer ring/disk becomes dominant, because the total amount of gas involved in the outer ring/disk is much larger than that in the central region, simply for the disk's surface-area effect. Hence, the ring/disk component becomes the major source for the total line profile.

Although the HI gas is distributed in a broader disk than CO, its densest region is close to the CO disk. Namely, the densest HI disk appears at 3–10 kpc for NGC 891 and at 5–10 kpc for NGC 4565. Moreover, the rotation curve attains its broad (nearly flat) maximum at these radii. Hence the CO and HI double horns arise from almost the same radii, at which the rotation curve is nearly flat.

Figure 4.36 shows a detailed comparison of CO and HI profiles for the edge-on spiral galaxy NGC 891, exhibiting almost identical profiles between CO and

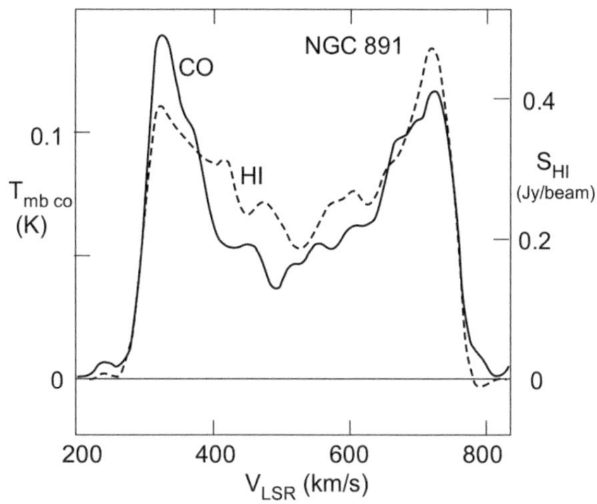

Fig. 4.36 Total line profiles of the $^{12}CO(J = 1 - 0)$ line emission (*full line*), and HI (*dashed line*) for the edge-on NGC 891 galaxies

HI. Such similarity between the two emission profiles has been confirmed for a number of spiral galaxies. This fact provides a basic step toward establishing the CO Tully-Fisher relation not only for nearby galaxies, but also for distant galaxies by observing CO lines using large millimeter and submillimeter facilities. Even though many of the sample galaxies have peculiar features such as active nuclei or interaction with a companion, there is a simple correlation between CO and HI.

4.10.3 The CO Tully-Fisher Relation: mm-Wave Cosmology

Beyond distances farther than a few hundred Mpc, angular resolutions of a few arc-minutes in HI observations become too large to resolve individual galaxies in a cluster. Interferometers such as VLA are not useful for the purpose because of the limited number of spectral channels (velocity resolution) caused by the limited number of autocorrelator channels. Furthermore, more red-shifted HI frequency results in increases in beam size as well as in interferences, which also makes resolution of distant cluster galaxies difficult (Figs. 4.37 and 4.38).

On the other hand, CO facilities have much sharper beams (e.g., 15 arc-sec with the Nobeyama 45-m telescope), with the use of which we would be able to resolve individual member galaxies in a cluster more easily, making it possible to avoid contamination by other member galaxies in a beam. Moreover, the larger the red-shift of an object is, the lower the CO frequency becomes, which results in a decrease in the system noise temperature because of atmospheric O_2 emission near 115 GHz: the more distant a galaxy is, the lower the noise temperature

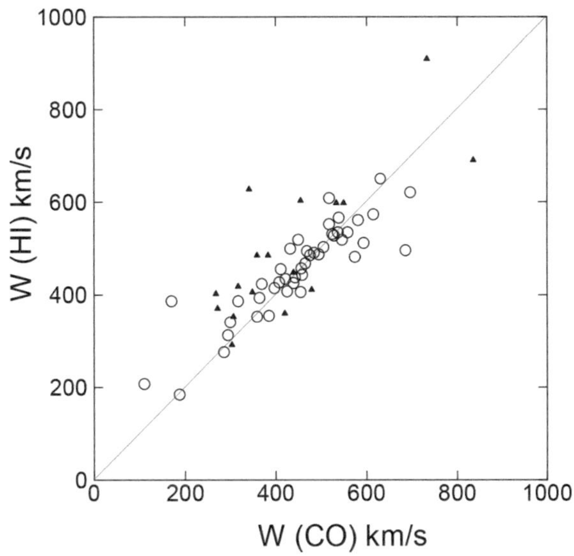

Fig. 4.37 CO versus HI line width corrected for inclination, for normal galaxies (*filled circles*) and interacting galaxies (*triangles*) [20].

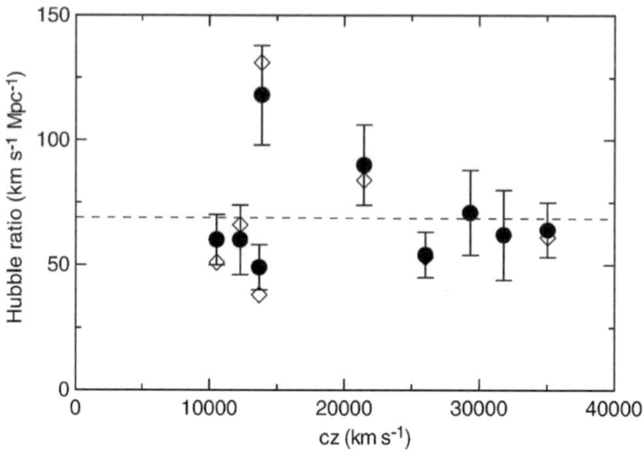

Fig. 4.38 Hubble constants obtained by CO Tully-Fisher relation [20]. The dashed line is the most recent value $H_0 = 69.32 \pm 0.80\,\mathrm{km\,s^{-1}\,Mpc^{-1}}$ [1]

becomes. The only disadvantage of the use of the CO line would be its sensitivity, particularly for distant galaxies. Actually, we need a few mK rms data for line-width measurements with a velocity resolution of $10\,\mathrm{km\,s^{-1}}$ for normal galaxies beyond $cz \sim 10{,}000\,\mathrm{km\,s^{-1}}$, for which an integration time of about 10 or more hours is required, and such observations are possible only by a long-term project with the largest mm-telescopes. However, although noise temperatures of current CO receivers, which are some hundred K in the present state, are still worse than those used in HI observations, we have the hope that they will be much improved in the near future to several tens K, which may result in an increase of the sensitivity by one or two orders of magnitude.

Because the distribution of HI gas is broad in a galaxy, HI line profiles are easily disturbed by interactions among galaxies, which is inevitable in the central region of a cluster. Such a disturbance might cause uncertainty in the HI line profiles for the Tully-Fisher relation. On the other hand, CO gases are more tightly correlated with the stellar distribution, and are less affected by tidal interaction. Of course, CO gas is distributed enough to a radius of several to 10 kpc, therefore the integrated line profiles manifest the maximum velocity part of the rotation curve.

References

1. Bennett, C.L., Larson, D., Weiland, J.L., et al.: Nine-year Wilkinson Microwave Anisotropy Probe (WMAP) observations: Final maps and results. ApJS **208**, 20 (2013). [Hubble constant WMAP9y]
2. Binney, J. Tremaine, S.: Galactic Dynamics, p. 146. Princeton University Press, Princeton (1987). [Galactic dynamics]

3. Dame, T.M., Hartman, D., Thaddeus, P.: The Milky Way in molecular clouds: a new complete CO survey. ApJ **547**, 792 (2001). [CO in the Galaxy]
4. Haslam, C.G.T., Salter, C.J., Stoffel, H., & Wilson, W.E.: A 408 MHz all-sky continuum survey. II – the atlas of contour maps. AA Suppl. **47**, 1 (1982)
5. Hou, L.G., Han, J.L.: The observed spiral structure of the Milky Way. AA **569**, 125 (2014). [HII regions]
6. Honma, M., Nagayama, T., & Sakai, N.: Determining dynamical parameters of the Milky Way Galaxy based on high-accuracy radio astrometry. PASJ **67**, 70 (2015). [Galactic constants by VERA]
7. Kalberla, P.M.W., Burton, W.B., Hartmann, D., et al. The Leiden/Argentine/Bonn (LAB) Survey of Galactic HI. Final data release of the combined LDS and IAR surveys with improved stray-radiation corrections. AA **440**, 775 (2005). [All sky HI]
8. Matsunaga, N., Kawadu, T., Nishiyama, S., et al.: A near-infrared survey of Miras and the distance to the Galactic Centre. MNRAS **399**, 1709 (2009)
9. Miyamoto, M., Nagai, R.: Three-dimensional models for the distribution of mass in galaxies. PASJ **27**, 533 (1975). [Galaxy potential]
10. Nakanishi, H., Sofue, Y.: Three-dimensional distribution of the ISM in the Milky Way Galaxy: II. The molecular gas disk . PASJ **58**, 847 (2006). [MW HI and CO 3D map]
11. Nakanishi, H., Sofue, Y.: Three-dimensional distribution of the ISM in the Milky Way Galaxy. IV. 3D molecular fraction and Galactic-scale H I-to-H2 transition. PASJ **68**, 5 (2016). [ibid]
12. Navarro, J.F., Frenk, C.S., White, S.D.M.: A universal density profile from hierarchical clustering. ApJ **490**, 493 (1997). [Dark halo potential]
13. Oka, T., Hasegawa, T., Sato, F., Tsuboi, M., Miyazaki, A.: A large-scale CO survey of the galactic center. ApJS **118**, 455 (1998). [CO in the GC]
14. Oort, J.H., Kerr, F.T., Westerhout, G.: The galactic system as a spiral nebula (Council Note). MNRAS **118**, 379 (1958). [HI view of the MW]
15. Pavlović, M.Z., Urošević, D., Vukotić, B., Arbutina, B., Göker, Ü.D.: The radio surface-brightness-to-diameter relation for galactic supernova remnants: sample selection and robust analysis with various fitting offsets. ApJS **204**, 4 (2013)
16. Sofue, Y.: The mass distribution and rotation curve in the galaxy. In: Gilmore, G. (ed.) Planets, Stars and Stellar Systems, vol. 5, chap. 19. Springer, Dordrecht (2013). [Rotation curves]
17. Sofue, Y.: Rotation and mass in the Milky Way and spiral galaxies. PASJ **69**, 1 (2017). https://arxiv.org/abs/1608.08350. [Review of RC and Mass in MW and galaxies]
18. Sofue, Y., Rubin, V.C.: Rotation curves of spiral galaxies. ARAA **39**, 137 (2001). [RC review]
19. Tully B., Fisher J.R.: A new method of determining distances to galaxies. AA **64**, 661 (1977). [Tully-Fisher relation]
20. Tutui, Y., Sofue, Y.: Effects of galaxy interaction on the Tully-Fisher relation: CO VS HI linewidths. AA **326**, 915 (1997). [CO TFR]

Chapter 5
The Galactic Center and Activity

5.1 Radio Continuum Features

Radio continuum features in the galactic center (GC) region are a superposition of various active radio sources, as illustrated in Fig. 5.1: star-forming regions distributed in a thin thermal disk, vertical structures that are mostly nonthermal closely related to poloidal magnetic fields, and the galactic center known as the radio source Sgr A, nesting a supermassive black hole.

5.1.1 Radio Maps

The radio emission is therefore a mixture of thermal and nonthermal emissions. In this chapter, we discuss the properties of the radio emission, reviewing the methods to distinguish thermal and nonthermal emissions. We then summarize various exotic structures, which are mostly perpendicular to the galactic plane, and discuss them in relation to vertical magnetic fields and to the manifestation of energy release from the nuclear disk. We comment on the similarity of large-scale ejection features to some radio bubbles and outflow features in external galaxies. Vertical cylinder structures of expanding rings of molecular gas are also discussed.

Figure 5.2 shows a wide-view radio continuum map around the galactic center at 408 MHz, exhibiting the bright galactic disk and numerous ridgelike structures extending toward high galactic latitudes. Figure 5.3 is a closer view of the galactic center region at 10 GHz, showing numerous radio sources. The strongest source is the nucleus of the galaxy, called Sgr A. The other sources near the galactic plane are thermal sources such as HII regions and nonthermal sources such as supernova

The original version of this chapter was published without printing chapter abstract. The abstracts for each chapter are updated in the front matter in the current version of the book. An erratum to this chapter can be found at https://doi.org/10.1007/978-981-10-3445-9_7.

Y. Sofue, *Galactic Radio Astronomy*, Lecture Notes in Physics 935,
DOI 10.1007/978-981-10-3445-9_5

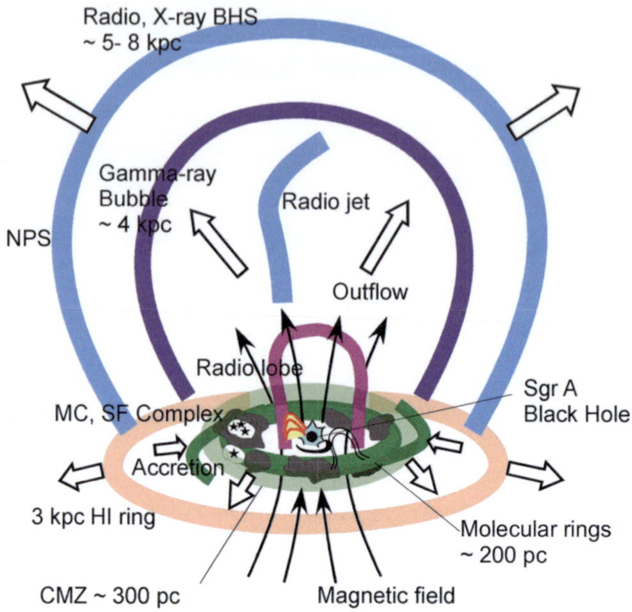

Fig. 5.1 Various features around the galactic center

remnants. Also evident are the vertical ejection features perpendicular to the galactic plane. Figure 5.4 shows an enlarged map, exhibiting these sources more clearly.

5.1.2 *Radio Spectrum*

The radio continuum emission from the GC region of $\sim 3° \times 3°$ is a mixture of nonthermal (synchrotron) and thermal (free-free) emissions (Fig. 5.4). In order to discriminate their emission mechanisms, it is essential to investigate their spectral characteristics. We describe the methods to separate the two major components of thermal and nonthermal (synchrotron) emissions. The conventional method to investigate the emission mechanism is to study the spectral index. Thermal emission shows a flat spectrum of $\alpha = -0.1$ with $S \propto \nu^{\alpha}$, whereas nonthermal emission usually shows a steeper spectrum.

Figure 5.5 shows the distribution of the radio spectrum obtained by using maps at 2.7 and 10 GHz, which shows generally a flat spectral index. Because of the spectral flatness, it was thought that the extended radio component comprises an ellipsoidal disk of diffuse thermal gas of a size $\sim 300 \times 100$ pc. However, later polarization observations revealed that the emission is mostly synchrotron, and hence the flat spectra are due to the flatter energy distribution of the emitting cosmic-ray electrons. This means that cosmic rays are accelerated in the wide area around the GC.

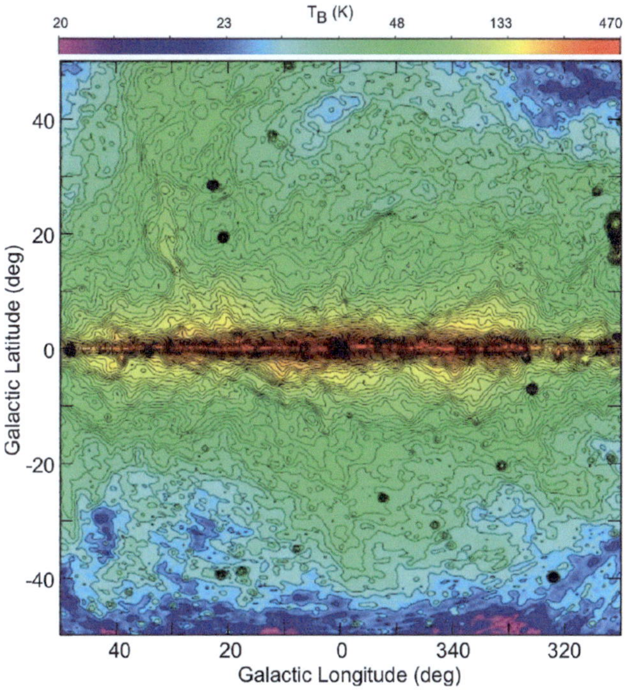

Fig. 5.2 Galactic center region at 408 MHz (Bonn-Parkes All Sky Survey: [3]), showing various ejection features from the central region

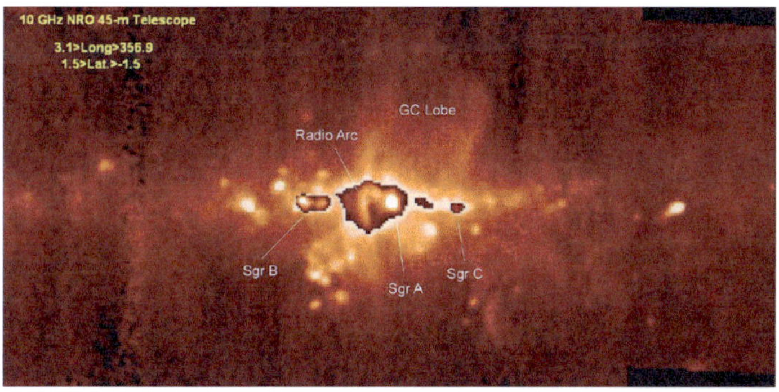

Fig. 5.3 10 GHz map of the central 6° × 3° region of the galaxy observed with the Nobeyama 45-m telescope

More detailed spectral index studies of radio data at larger frequency separation from 845 MHz to 43 GHz have shown that the spectral indices in the strongly polarized radio arc and bridge region near Sgr A are also flat or even inverted (positive). Flat spectra are thus obtained widely in the galactic center even where strong linear polarization was detected.

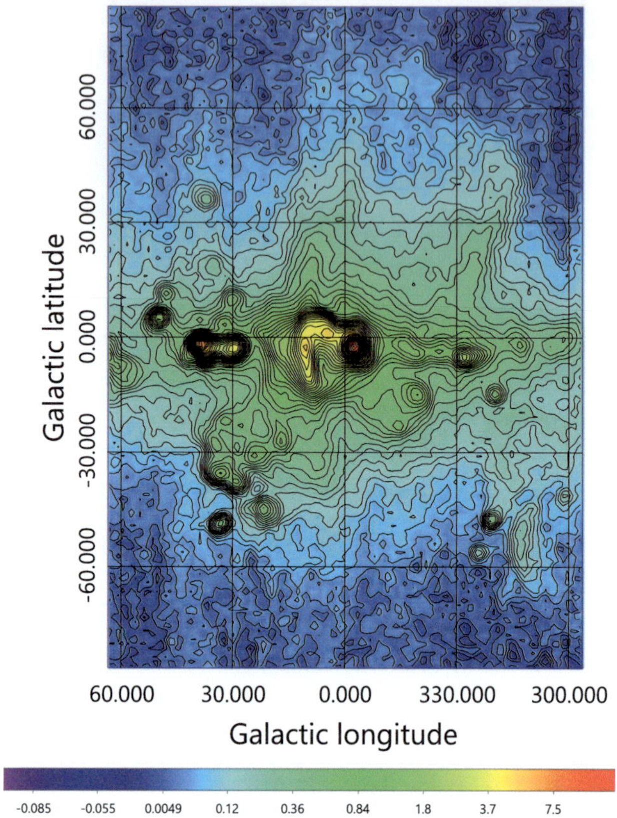

Fig. 5.4 Same as Fig. 5.3, but for the central $2 \times 3°$ region. *Color bar* indicates brightness temperature T_B in K. Contours are drawn logarithmically at equal dex (δ log T_B) interval

Hence the flat spectra observed near the galactic center are no longer taken as an indicator of thermal emission. We note that many active galactic nuclei in extragalactic systems also have flat radio spectra despite their nonthermal characteristics. However, the questions of why the spectra are so flat, or how high-energy electrons are supplied so efficiently in such a wide area of the galactic center, as yet remain unanswered.

5.1.3 Linear Polarization and Magnetic Field

A direct and more convincing way to distinguish synchrotron radiation is to measure the linear polarization. However, an extremely high Faraday rotation caused by the dense interstellar matter and long path length through the galactic plane depolarizes the emission by the finite-beam and finite-bandwidth effects, which makes the

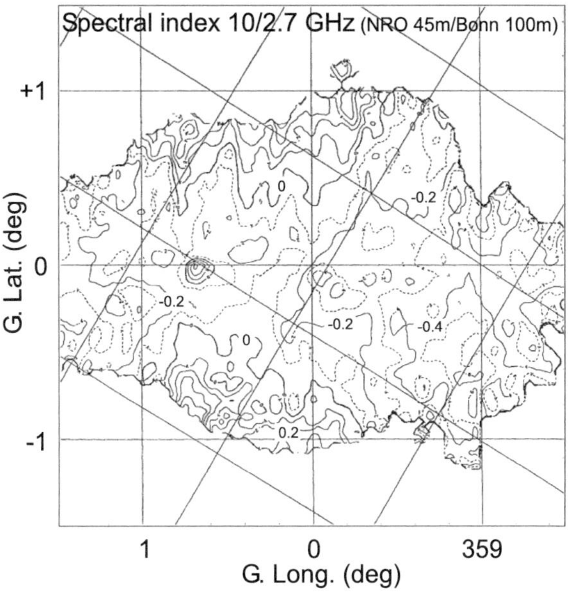

Fig. 5.5 Radio spectrum distribution in the central 3° region of the galaxy as observed at 10 and 2.7 GHz

measurement difficult. This difficulty has been resolved by the development of a multifrequency, narrow-band Faraday polarimeter as well as by high-resolution and high-frequency observations using the very large array (VLA). Very large rotation measure ($RM >\sim 10^3$ rad m^{-2}) and high-degree (10–50%) polarization have been observed along the radio arc and on the eastern ridge of the galactic center lobe.

At high frequencies the Faraday depolarization becomes less effective. Figure 5.6 shows strongly polarized regions in the GC. The Faraday rotation measure and distribution of intrinsic magnetic field directions are also shown. Linear polarization as high as $p \sim 50\ \%$ along the radio arc at mm wavelengths is detected. This high polarization is nearly equal to the theoretical maximum, $p_{\max} = (\alpha + 1)/(\alpha + 7/3) \simeq 47\%$ for the arc region where the spectral index has the value of $\alpha \simeq +0.2$. This fact implies that the magnetic field is almost perfectly ordered. This is also consistent with the VLA observations showing straight filaments suggestive of a highly ordered magnetic field. From these observations of linear polarization it is clear that the radio emission near the galactic center, in particular around the radio arc, is nonthermal despite the flat or inverted spectra.

5.2 Central Disk and Star Formation

The nuclear disk comprises a high-density rotating disk of thickness ~50 pc and radius ~200 pc. The disk consists of numerous clumps of thermal emission regions, most of which are active star-forming regions and HII regions detected in the

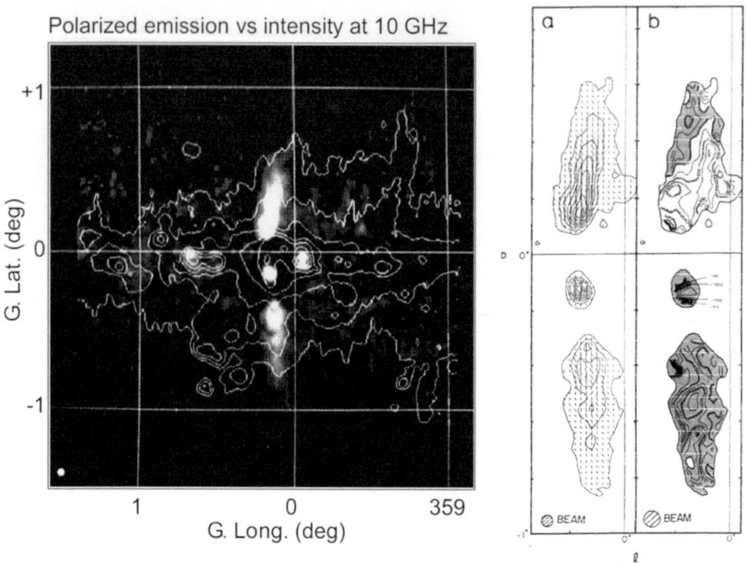

Fig. 5.6 (*Left*) Linearly polarized region at 10 GHz of the GC (*grey plumes*) superposed on intensity contours. (*Right*) Magnetic field directions and Faraday rotation measure on a enlarged map near the plumes [16]

hydrogen recombination lines. Typical HII regions are named Sgr B, C, D, and E. The total HII mass of $2 \times 10^6 M_\odot$ has been estimated, and the production rate of Ly continuum photons of $3 \times 10^{52}\,\mathrm{s}^{-1}$ is required to maintain this amount of HII gas. The star-forming rate, which is assumed to be proportional to the Ly continuum photon flux, of the central few hundred pc region reaches almost 10% of the total star-forming rate of the galaxy. The HII regions are surrounded by expanding shells and cylinders of molecular gas.

Because of the high concentration of molecular gas in the central few hundred pc region, the star-formation efficiency (*SFR*/molecular gas mass) in the central region is not high as in the normal SF regions. Or, in turn, the ratio (γ) of the gaseous mass to the SF rate can be taken as the "youth" of the SF region, and the value of γ is large compared to HII regions in the outer disk. The central SF region, particularly an SF region such as Sgr B2, is "younger" in the sense that the "fuel" gas is overabundant compared to newly born stars.

The galactic center region has also been observed intensively at various molecular lines. In particular, CO line observations have shown a global distribution of the molecular gas and have revealed a high concentration in the central few hundred pc. This fact clearly indicates that intense star formation in the central region occurs deeply coupled with the wealth of molecular gas. The intensity distributions of the radio continuum and molecular line emissions are well correlated near the galactic plane, and significantly asymmetric with respect to the galactic center (Sgr A). They are both stronger on the positive longitude side than in the negative longitude regions.

5.2.1 *Sgr B: A Molecular Complex with Star-Forming Region*

The radio source Sgr B2 consists of compact HII regions with strong thermal emission with recombination lines at a velocity of $V_{\rm LSR} = 65\,{\rm km\,s^{-1}}$. A dense molecular cloud with an angular size of about 0°.05 at $V_{\rm LSR} = 60 \sim 70\,{\rm km\,s^{-1}}$ is associated not only with velocity, but also in position with Sgr B2. The continuum radio emission suffers from strong H_2CO molecular line absorption at $V_{\rm LSR} = 62.5\,{\rm km\,s^{-1}}$. This indicates that the molecular cloud lies in front of the continuum source, and that probably Sgr B2 (HII region) and the molecular cloud are in contact with each other along the line of sight. Namely, star formation in Sgr B2 may be taking place at the farther end of the molecular cloud along the line of sight.

The Sgr B2 molecular cloud is a part of an even larger molecular complex; at a similar velocity, 60–70 $\rm km\,s^{-1}$, the cloud is embedded in a wide outskirt with a diameter of approximately 0°.25. We call this complex the "Sgr B2 molecular complex." The total mass of this complex can be estimated using the integrated intensity and the extent as read from the ^{13}CO maps. We use the relation $\sigma = 32I$, with σ being the surface mass density of molecular gas in $M_\odot\,{\rm pc^{-2}}$ and I being the ^{13}CO line intensity in $\rm K\,km\,s^{-1}$. The total mass of the Sgr B2 molecular complex, including its outskirts, is estimated to be $1.5 \times 10^6\,M_\odot$. The thermal source Sgr B1 lies close to Sgr B2. At the same velocity as Sgr B1 ($45\,{\rm km\,s^{-1}}$) there exists a molecular complex that extends for an angular extent of $\Delta l \times \Delta b \simeq$ 0°.4 × 0°.3. It is remarkable that this molecular complex makes a large loop with its clearest appearance at $V_{\rm LSR} = 30 \sim 40\,{\rm km\,s^{-1}}$.

This loop feature has a systemic velocity of about $40\,{\rm km\,s^{-1}}$ and can be recognized over a wide velocity range from $V_{\rm LSR} = 70\,{\rm km\,s^{-1}}$ to $V_{\rm LSR} \simeq 20\,{\rm km\,s^{-1}}$. We emphasize that the recombination-line velocity ($V_{\rm LSR} = 45\,{\rm km\,s^{-1}}$) of Sgr B1 coincides with the velocity of this molecular loop, which indicates that the loop is physically related to Sgr B1, possibly driven by the activity of this HII region. From these facts we may interpret that the loop feature is a molecular shell surrounding Sgr B1; we call it the "Sgr B1 molecular shell."

The large velocity dispersion of the shell ($V_{\rm LSR} = 20 \sim 70\,{\rm km\,s^{-1}}$) suggests that it is expanding at a velocity of about $25\,{\rm km\,s^{-1}}$ referring to its center velocity. The size of this shell is about 40×30 pc for a distance of 8 kpc. If the expanding velocity is $25\,{\rm km\,s^{-1}}$, the age of the shell would be on the order of 10^6 yr. Hence, the expansion of the shell may have started about 10^6 yr ago when Sgr B1 was still more active than today. The total mass of the shell is estimated to be $2 \times 10^6\,M_\odot$. This leads to a kinetic energy of the expanding motion of about 1.3×10^{52} ergs.

The shells and rings are probably driven by activities in the star-forming regions, such as stellar winds and/or supernova shocks. In fact, the kinetic energies of the shells, $E \sim 10^{50-52}$ ergs, can be supplied by several supernovae. According to the starburst-ring formation scenario, successive SN explosions and stellar winds cause high pressure in the central region of a dense nuclear molecular disk, and a cavity forms near the center. Then, the upper and lower parts of the disk are blown

Table 5.1 Radio continuum sources and molecular clouds in the galactic center region

Source	Coord. G(l, b)	Property ($\mathrm{km\,s^{-1}}$)	V_{LSR} ($\mathrm{km\,s^{-1}}$)	$V_{\mathrm{exp/rot}}$ ($M_\odot$)	M_{MC} (erg)	E_{kin}
Sgr A	–0.05–0.05	Nucleus	0	$\sim \pm 100$	$\sim 10^4$	–
Sgr B1	0.52–0.05	MC SNR	$\sim +35$	25	$\sim 10^{5-6}$	$\sim 10^{52}$
Sgr B2	0.66–0.04	MC HII	$\sim +65$	...	$\sim 10^6$	...
Sgr C	G–.56-.08	MC HII	~ -60	...	$\sim 10^5$	...

off while forming a cylindrical outflow perpendicular to the disk. Accordingly an expanding molecular ring forms within the disk. It is possible that similar cavity-ring formation takes place within a single massive cloud near the galactic center triggered by interior star formation or miniburst: namely minirings/shells can be formed by minibursts associated with individual clouds.

The HII region Sgr B2 may still be embedded within or in contact with a dense molecular complex, and star formation (or a miniburst) in Sgr B2 has possibly been triggered by the expanding Sgr B1 shell. The molecular shell centered by Sgr B1 is in a cavity phase and expanding. The molecular ring around G–0.27 is in a late ring phase, in which the upper and lower parts of the cloud had already been blown off. It is thus possible that the G–0.27 ring is the oldest, the Sgr B1 shell is moderate, and the Sgr B2 complex the youngest star-forming (miniburst) regions. The fact that they are not necessarily located close to each other suggests that starburst sites are wandering around the galactic center.

Table 5.1 summarizes the position and velocity relationship between the molecular and HII features in the galactic center region.

5.2.1.1 Sgr C

On the integrated ^{13}CO map the continuum source Sgr C lies near a position where the molecular emission has a depression. On the channel maps of the CS($J = 2-1$) line emission we find a complex of molecular gas at $V_{\mathrm{LSR}} = -40 \sim -60\,\mathrm{km\,s^{-1}}$ that although associated with Sgr C, is position ally shifted toward the southeast by about $5'$. Again we notice that the continuum source does not position ally coincide with the peak position of its associated molecular complex.

The Sgr C molecular complex extends for approximately 35 pc not only along the galactic plane but also in the direction perpendicular to the plane. In particular, it has a negative-latitude extension at $V_{\mathrm{LSR}} = -50 \sim -60\,\mathrm{km\,s^{-1}}$. Another molecular feature at high negative velocity, $V_{\mathrm{LSR}} = -100 \sim -140\,\mathrm{km\,s^{-1}}$, is associated with this region. This high-negative velocity feature comprises two symmetrical plumes extending from G–0.50–0.05 vertically to the galactic plane about $\pm 0^\circ.2$. This might be associated with possible high-velocity ejection along the galactic center lobe.

5.3 Molecular Ring

Figure 5.7 shows the total ^{12}CO line intensity map integrated over the full range of the velocity ($-200 \le V_{\rm LSR} \le 200\,{\rm km\,s^{-1}}$) as obtained by integrating the CO survey data of the galactic center region. The intensity map can be used to obtain the column density of the molecular gas by multiplying the integrated intensity by the CO-to-H_2 conversion factor $X_{\rm CO}(R = 0) = 0.92(\pm 0.2) \times 10^{20}$ ($H_2\,{\rm cm^{-2}/K\,kms\,s^{-1}}$). The distribution of the column density may then be integrated in the area around the galactic center at $-1° \le l \le 1°$ (within radius 140 pc), yielding the total molecular mass of $M_{H_2} = 5 \pm 1 \times 10^7\,M_\odot$.

Figure 5.8 shows the longitude-velocity (L, V) diagram corresponding to Fig. 5.7. The figure reveals various features, among which the following structures are prominent.

- Tilted straight ridge at velocities $|V_{\rm LSR}| \sim -150 - +150\,{\rm km\,s^{-1}}$ showing a rigid-body rotation, which represents a rotating dense molecular ring with radius of 120 pc
- Widely extending broad oval feature in the (L, V) plane, or a parallelogram at $|l| \sim\le 1.2°$ and $|V_{\rm LSR}| = -200 - +200\,{\rm km\,s^{-1}}$ having noncircular velocity of $\pm 100\,{\rm km\,s^{-1}}$ at $l \sim 0°$

The rigid-body component has a mass $4 \times 10^7\,M_\odot$, about 80% of the total mass in $|l| \le 1°$. On the other hand, the expanding ring (or the parallelogram) shares only $7 \times 10^6\,M_\odot$ (15%) in the region at $|l| < 1°$.

5.3.1 *120-pc Ring and Sgr B Molecular Complex*

In the (l, v) diagram the rigid-body feature is clearly traced by an arm from $l = 0°.9$ near the Sgr B complex toward negative longitude at $l = -1°.0$, which is called

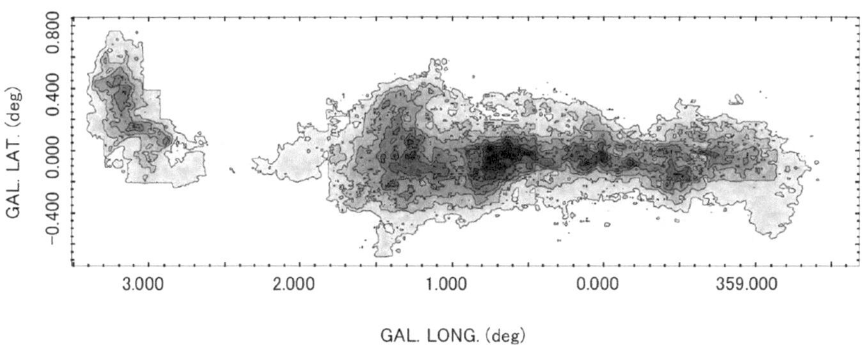

Fig. 5.7 ^{12}CO line intensity distribution in the galactic center [10]

Fig. 5.8 Longitude-velocity diagrams of the ^{12}CO line emission of the galactic center at $b = +0^{\circ}.02833$ and $-0^{\circ}.1888$ [10], and schematic LV diagram of the GC molecular arms, corresponding to the 120-pc molecular ring

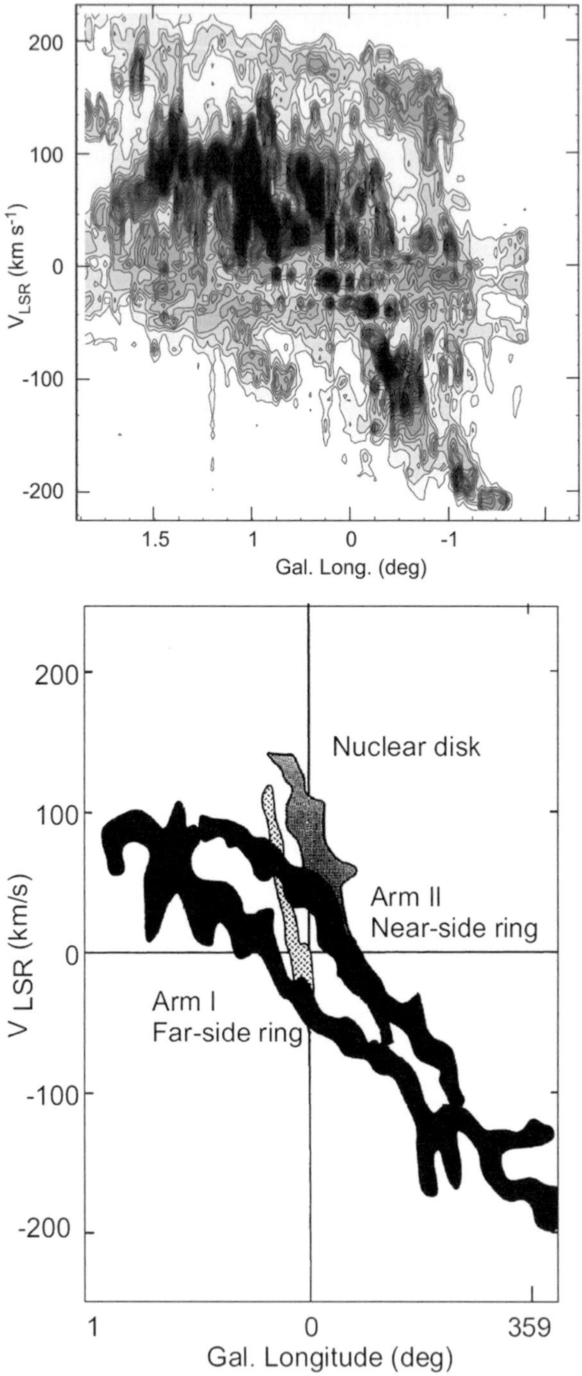

the galactic center Arm I. The angular extent is as long as $1^\circ.9$ (280 pc) in the longitudinal direction, whereas the thickness in the b direction is as thin as $\sim 5'$. The Sgr B molecular complex is much extended in the b direction by about $0^\circ.4$ (60 pc), and comprises a massive part of the arm. The mass of molecular gas involved in Arm I is $M \sim 2 \times 10^7 \, M_\odot$.

Another arm called Arm II is traced as a bright ridge from $l \sim 0^\circ.3$ to $-0^\circ.7$, and the thickness is about $6'$ (15 pc). The mass of Arm II is estimated to be $\sim 1.4 \times 10^7 \, M_\odot$. Thus, the total mass involved in Arms I and II is estimated to be $\sim 3 \times 10^7 \, M_\odot$, and shares almost 70% of the total gas mass in the observed region.

The thickness (minor axis length) is estimated to be $8'$ from the maximum separation between Arms I and II at $l \sim -0^\circ.2$. Therefore, the inclination of the I+II ring is $i = 85^\circ$ from the minor-to-major axis ratio. The center of the ring, as fitted by the above figures, is at (l,b) $=(0^\circ.1, 0^\circ.0)$. We call this ring the 120-pc molecular ring.

5.3.2 *Face-On View of the 120-pc Ring*

If the rotation of the ring is circular at velocity V, the (l, v) diagram can be transformed to the spatial distribution. The line-of-sight distance y of the element from the nodal line can be calculated by $y = \pm|x|\sqrt{\left(\frac{V}{v}\right)^2 - 1}$. Figure 5.9 shows a thus-obtained face-on map of the molecular gas for an assumed rotation velocity of $V = 150 \, \mathrm{km \, s^{-1}}$. In this view, Arms I and II comprise a global ring structure, which is tilted and slightly bent. Its angular extent in the major axis is $1^\circ.6$ from $l = -0^\circ.7$ to $0^\circ.9$, and thus, the major-axis length (diameter) is 240 pc, and the radius 120 pc. The connection of Arms I and II is not clear from this map because of the ambiguous position determination near the node, which arises from an unknown precise rotation curve.

5.4 Expanding Ring and Parallelogram

In addition to the 120-pc ring composed of Arms I and II traced as the tilted dense ridge in the (l, v) diagram, an oval feature appears composing an ellipse or a parallelogram in the (l, v) diagram. The feature is recognized as a broad elliptical distribution in the (l, v) diagram, and is called the (l, v) "parallelogram" or "ellipsoid." The (l, v) ellipsoid indicates the noncircular motion of the gas at radial velocities $v_r \sim \pm 100 \, \mathrm{km \, s^{-1}}$ at $l \sim 0^\circ$. There have been two interpretations about its origin.

One interpretation is that the feature is produced by an expanding molecular gas ring (EMR) driven by an explosive event or an outflow from the central region. Another model attributes the motion to noncircular flows of gas rotating in an oval gravitational potential produced by a stellar bar.

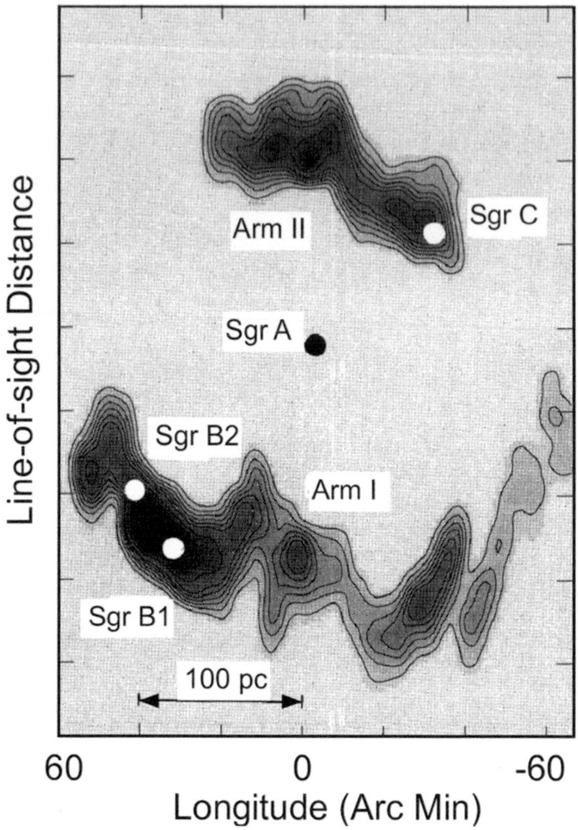

Fig. 5.9 120-pc molecular ring as seen from the *top* of the galaxy

In the EMR model the radial velocity is represented by

$$v = V_{\rm rot} + V_{\rm exp}\sqrt{1-(l/l_{\rm EMR})^2}$$

where $V_{\rm exp} \simeq 100\,{\rm km\,s^{-1}}$ is the expanding velocity and $l_{\rm EMR} \simeq 1^\circ.2$ ($\sim$200 pc) is the radius of the EMR. In the bar-potential model, the motion is a superposition of the circular rotation and an oval flow due to the bar potential.

5.5 Nuclear Supermassive Black Hole at Sgr A*

5.5.1 Massive Object in the Galactic Center

Sgr A* is the nucleus of the galaxy which is a compact radio source unresolved even with the use of very long baseline interferometer (VLBI). The radio flux shows

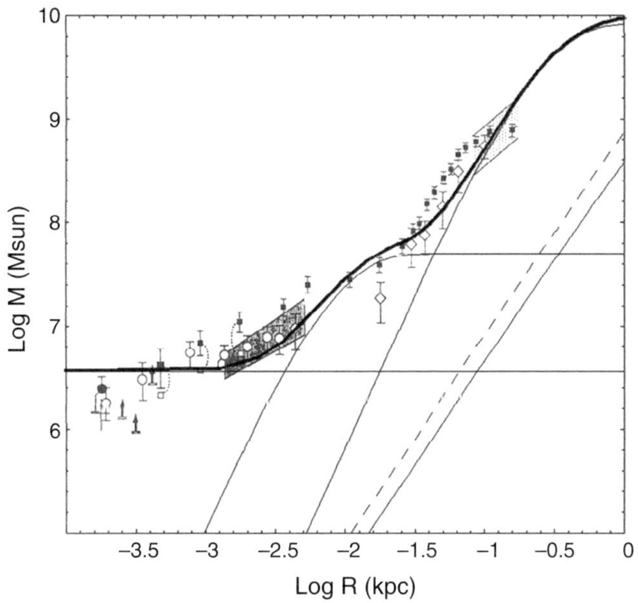

Fig. 5.10 Virial mass enclosed within radius R inferred from central rotation velocities (*solid lines*), as compared with those estimated from velocity dispersion of stellar objects by *dots* and *shade* [1]. The enclosed mass inside $R \leq 1$ pc tends to a constant at $\sim 3 \times 10^6 M_\odot$, indicating a point mass suggestive of a massive *black* hole.

variability with a short period, which indicates that the size, as on the order of period times the light speed, is supposed to be very compact, less than $\sim 10^9$ km ~ 10 AU.

Indirect evidence for a massive object has been obtained by measuring the mass distribution around the nucleus. From radial velocity dispersion of infrared stars and ionized gases, the Virial mass inside a few pc can be measured. The innermost rotation curve is also a powerful tool to map the mass. Figure 5.10 shows the enclosed mass within radius R around Sgr A* as measured from various radial velocity observations.

The total mass decreases with decreasing radius, obviously due to the decrease of the volume. If the central region has finite density distribution, the enclosed mass at radius $R \sim 0$ would become zero. However, the figure shows that the total mass tends to a constant value of $\sim 3 \times 10^6 M_\odot$ at $R < 1$ pc.

5.5.2 Sgr A*: A Supermassive Black Hole

More precise measurement of the mass around Sgr A* has been obtained by proper motion measurement of stellar objects. Proper motion observations of infrared sources, mostly asymptotic giant branch (AGB) stars, revealed that the stars are

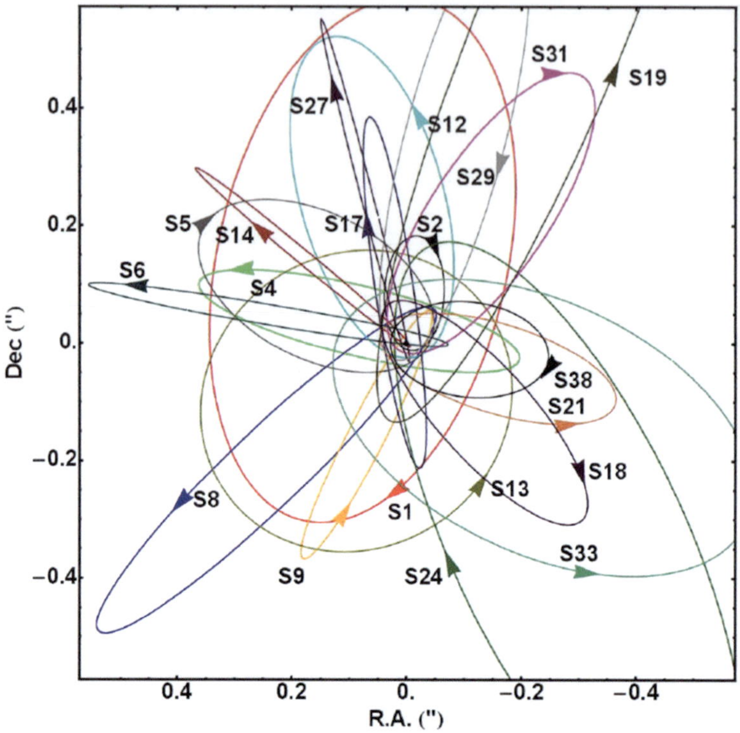

Fig. 5.11 Elliptical orbits of infrared stars around Sgr A* around the focal point coinciding with Sgr A*, indicating that there is a massive *black* hole at the position [2]

orbiting Sgr A* at velocities as high as $100 \sim 1500\,\mathrm{km\,s^{-1}}$. Their motions were found to draw closed elliptical orbits (Figs. 5.11 and 5.12).

From analysis of the elliptical motion on individual elliptical orbits, their focal points (centers of gravity) were found to coincide with the position of Sgr A*. The best-fit mass of the point was estimated to be $3.6 \times 10^6\, M_\odot$. The per-galactic distances of the orbits are as small as $\sim 0''.05$, or $\sim 0.002\,\mathrm{pc} \sim 400\,\mathrm{AU}$. Thus the central mass is enclosed within an extremely small volume of this radius, implying that the mean stellar density would be higher than $10^{14}\, M_\odot \mathrm{pc}^{-3}$.

5.5.3 *Black Holes in External Galaxies*

Supermassive black holes are universally observed in the nuclei of galaxies. Accumulating evidence for supermassive black holes has been obtained by high-resolution optical and radio-line spectroscopic observations. Particularly firm

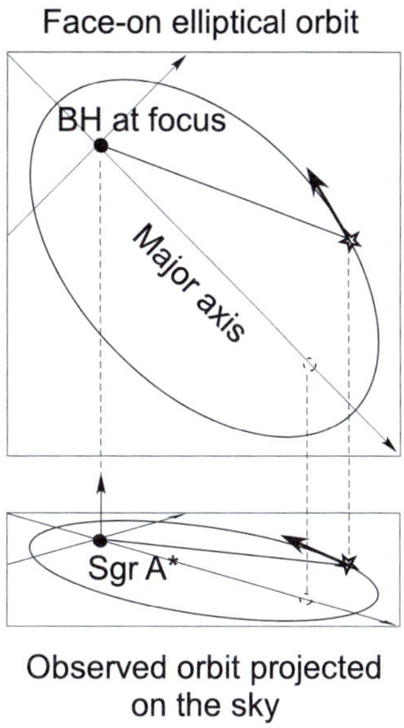

Fig. 5.12 Observed orbit is projected in the sky. The true elliptical orbit can be uniquely determined by tracing the star's loci on the orbit and the position of Sgr A* supposed as the position of BH. Then the mass of the BH is calculated by applying Kepler's law

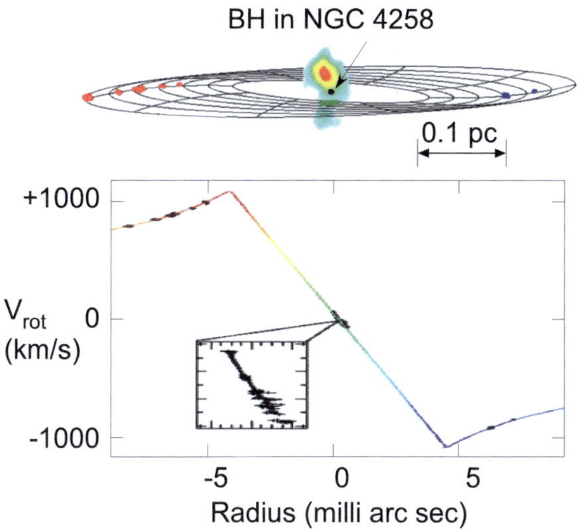

Fig. 5.13 Massive *black* hole of 30 million solar masses in the spiral galaxy NGC 4258, around which many maser sources are rotating in Keplerian orbits, as observed by VLBI at milli-arcsecond resolution [7]

evidence was obtained in the spiral galaxy NGC 4258 by VLBI measurements of circumnuclear water maser sources emitting extremely high-brightness temperature lines at 23 GHz.

Figure 5.13 shows a VLBI imaging of the maser sources around the nucleus of NGC 4258, which are aligned approximately on a straight line. The sources exhibit very high radial velocities at ~ -900, ~ 0 at $\sim +900\,\mathrm{km\,s^{-1}}$ with respect to the systemic velocity of the galaxy. A velocity-to-displacement plot of the sources as shown in the lower panel indicates that the high-velocity components obey the Keplerian rotation curve, revealing that they are rotating around a pointlike mass. The low-velocity components show a rigid-body rotation as shown by the diagram inset.

From these velocity characteristics, it is argued that the sources are parts of a Keplerian rotation disk, on which the sources are rotating and emitting the maser line in the directions either parallel or perpendicular to the velocity vectors. The parallel emission is observed at the tangential points as the high-velocity components as blue- and red-shifted lines. The perpendicular emission is observed at their meridian passage.

Using the rotation velocity $V_{rot} = 1000\,\mathrm{km\,s^{-1}}$ at $R = 0.1\,\mathrm{pc}$ for the galaxy's distance of 8 Mpc, the central mass is estimated to be $M_\bullet \sim 4 \times 10^7\,M_\odot$. This is about an order of magnitude more massive than the black hole in our galactic center.

5.6 Circumnuclear Activities

5.6.1 *Sgr A Halo and Mini Spirals*

The nucleus is surrounded by a nebulosity that exhibits three-armed peculiar spirals of a size ∼3 pc, showing the thermal radio spectrum. From infrared line spectra, the minispirals show a continuous velocity gradient along each arm with radial velocities ranging from ~ -100 to $+100\,\mathrm{km\,s^{-1}}$. The velocities and spiral shapes are fitted by three ellipsoids with Sgr A* at the focal position, indicating that the spirals are the loci of orbiting gaseous oval rings. This configuration has been interpreted as representing the place of gaseous accretion from a tilted ring of molecular gas of radius ∼3 pc, which is rotating at a velocity of about $100\,\mathrm{km\,s^{-1}}$. The region including all these features is called Sgr A West, which shows a flat spectrum in radio.

A shell type radio continuum structure is superposed on the eastern edge of these spirals, and is called Sgr A East. The feature is supposed to be a supernova remnant from its steep spectrum indicating a nonthermal emission.

All these features are further embedded in an extended radio halo of radius about 10 pc. The radio-emitting region containing all these radio sources is called Sgr A (Fig. 5.14).

5.6.2 *Thermal Filaments*

Complex filaments extend from Sgr A toward the north, composing a bridge between Sgr A and the straight radio arc described in the next subsection (Fig. 5.15). Detection of hydrogen recombination lines and association of molecular gas at negative velocities indicate its thermal characteristics, and the structures are called the thermal arched filaments.

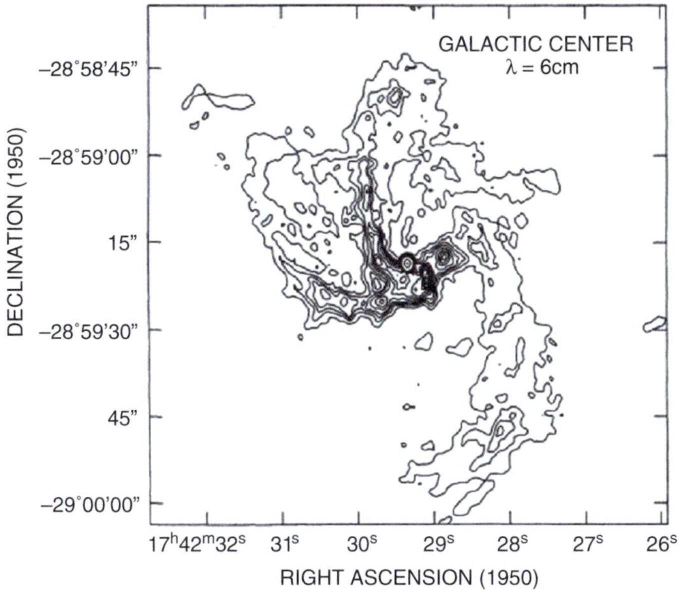

Fig. 5.14 Sgr A* and the three-armed minispirals at 6 cm [4]

Fig. 5.15 Sgr A with the halo and the radio arc [6, 17]

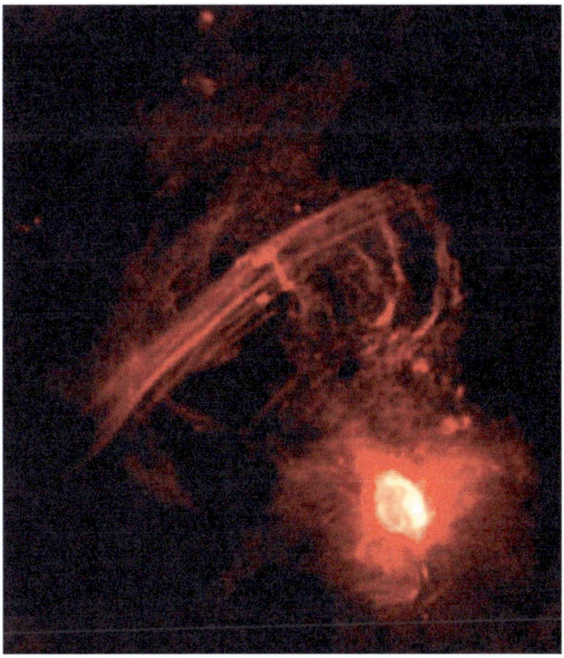

In spite of thermal radio emission, a large Faraday rotation has been detected toward the bridge, which indicates the existence of a magnetic field along the thermal filaments. From these facts the bridge is considered to be a magneto-ionic structure connecting the radio arc with Sgr A along which the gas is flowing toward the center and accretes to the nuclear molecular ring.

The complex appearance in the bridge suggests turbulent motion inside the bridge. Velocity dispersion as high as 30–50 km s^{-1} has been observed in the bridge from recombination-line observations. The velocity dispersion gets largest near the arc with 60–70 km s^{-1}, indicating dynamical interaction of the bridge with the arc.

5.6.3 Radio Arc and Vertical Magnetic Fields

The radio arc was originally found in radio continuum survey maps of the galactic plane using single-dish telescopes. By interferometric high-resolution observations such as using the VLA, the arc was resolved into many straight filaments (Fig. 5.15). The straight filaments run perpendicularly to the galactic plane, and extend more than $\sim$100 pc toward positive latitudes. The midpoint of the arc is strongly polarized ($\sim$20–50%), and shows a high Faraday rotation ($RM \sim$ a few $10^3 - 10^4$ rad m^{-2}).

The theoretically maximum polarization as high as $\sim$50% along the arc indicates a highly ordered magnetic field and is consistent with the VLA straight filamentary appearance. The magnetic field direction, as determined from the intrinsic polarization angles, is parallel to the filaments and, hence, vertical to the galactic plane. Field strength is estimated to be as high as $\sim$1 mG from the Faraday rotation measure as well as the energy equipartition analysis of cosmic rays and magnetic fields.

Interferometric observations of the filaments in the arc at 43 GHz revealed that some filaments are not visible at this high frequency. Assuming the field strength of the order of 1 mG, the lifetime of cosmic-ray electrons emitting at 43 GHz is estimated to be about 4000 yr. This implies that the filaments of a strong magnetic field may be a time-variable and transient feature, being temporarily illuminated by recently accelerated high-energy electrons.

The higher latitude extensions of the arc, both toward positive and negative latitudes for more than 100 pc, are also polarized and are called polarized plumes. The degree of polarization reaches as high as 20% at 10 GHz. The magnetic field directions are parallel to the radio arc, or vertical to the galactic plane.

The sense of rotation measure reverses from positive to negative latitude sides, indicating a reversal of the line-of-sight component of the magnetic field above and below the galactic plane. The field strength is estimated to be 10–100 μG.

In addition to the filaments in arcs, numerous straight filaments, called "threads" are observed at low radio frequencies. They appear to be distributed rather independently of the other major radio sources. They are roughly perpendicular to the galactic plane. From their thin and straight appearance the threads are likely magnetic structures, and their vertical nature is consistent with a large-scale poloidal field in the central region.

5.7 Galactic Center Lobe

The galactic center radio lobe (GCL) is a loop or a two-horn structure of about 200 pc diameter extending vertically over the galactic plane (Figs. 5.4 and 5.16). The cross-section of the lobe parallel to the galactic plane shows a double-horn intensity distribution, indicating a cylindrical structure. The eastern ridge of the lobe is an extension from the radio arc and is strongly polarized. The magnetic field is shown to run parallel to the ridge.

The ridge also extends toward the negative latitude, where a symmetric polarized plume is found. The western ridge emerges from Sgr C and is clearly the extension of the VLA filament. The detection of the filament and polarization in the western ridge indicates the existence of a magnetic field. Recently, a complex of molecular gas and warm dust associated with the western ridge of the lobe has been detected, and a velocity jump in the CO-line spectra with a shock front coinciding with the western ridge has been found.

Formation of the lobe structure has been modeled in various ways: an explosion hypothesis requires an energy injection near the nucleus, which produces an expanding shell into the halo. An MHD acceleration model has also been proposed, in which the gas is accelerated by a twisted poloidal magnetic field by a gas disk accretion.

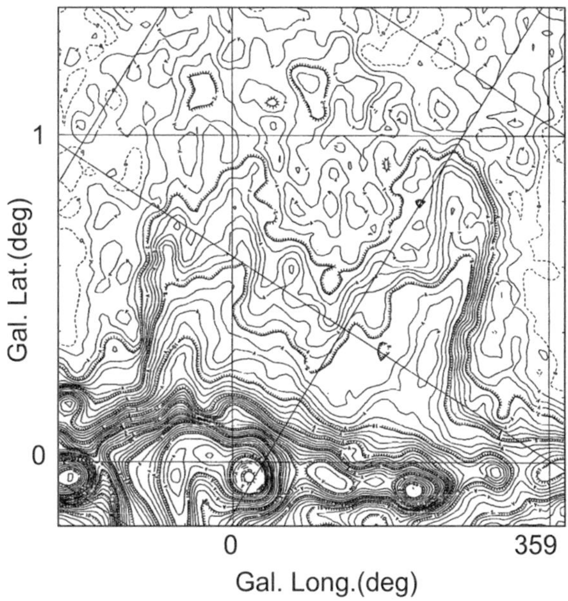

Fig. 5.16 Galactic center lobe (GCL) at 10 GHz

5.8 Vertical Structure

5.8.1 Poloidal Fields

Polarization observations of the galactic center region show that the field direction projected on the sky along the radio arc and the eastern ridge of the galactic center lobe is perpendicular to the galactic plane. The polarization is visible in a wider area of $1.5^\circ \times 1.5^\circ$ area around Sgr A, and shows that the magnetic field is widely distributed in the galactic center region. Measurements of the Faraday rotation show that the rotation measure (RM) reverses from the lower side of the galactic plane to the upper side, indicating reversal of the line-of-sight component of the field. The reversal of the RM is also observed across the rotation axis of the galaxy. From these, a poloidal field model has been proposed in which the field lines are twisted by the disk rotation. However, it is often claimed that the very straight nature of the filaments in the radio arc is suggestive of unbending field lines. If this is the case, the Faraday reversal should be attributed to intervening fields between the arc and the Sun. Because the RM reversal happens at $\sim 10'$-scale and the RM amplitude is as large as $RM \sim 10^4$ rad m^{-2}, it is reasonable to suppose that the Faraday effect occurs in the nuclear disk.

If we stand on the primordial origin hypothesis of galactic magnetic fields, we should inevitably consider the evolution of the vertical component of the primordial field. The timescale with which a magnetic field component perpendicular to the disk plane diffuses away from the galaxy may be estimated as $\tau_{\rm vert} \sim r^2/lv \sim 10^{11}$ y. Therefore, the vertical component trapped in intergalactic space can never escape from the galaxy, but is transferred to the central region when the galaxy contraction proceeds. A rough estimation shows that the field strength of the vertical component in the galaxy is proportional to the surface mass density, and a strong vertical field, some $10\,\mu$G, is expected to exist in the central 1-kpc region. On the other hand, the disk component parallel to the galactic plane is annihilated in the central region more efficiently than in the outer region and a rather weaker disk field is expected there.

This scenario of evolution of the galactic field thus predicts that a vertical field dominates in the central region, and the disk spiral field has the maximum in the outer disk several kpc away from the central region. This is consistent with the fact that our galaxy possesses a twisted poloidal field in the center, whereas the disk field (parallel to the galactic plane and spiral arms) is strongest at $r \sim 5$ kpc. Similar characteristics have been found in spiral galaxy M31: the central radio source shows polarization consistent with the vertical field. The projected field direction is perpendicular to the major axis, and the disk field, which is a superposition of ring and spiral, has the maximum at $r \sim 10$ kpc.

Given a large-scale poloidal field, which penetrates the rotating dense nuclear disk, a twist of the field is inevitable. The twist will then propagate toward the halo, accelerating the gas perpendicular to the disk plane. This will result in a cylindrical outflow of gas and cosmic rays, which will account for the formation of the galactic center lobe. The large-scale poloidal field in the halo will also be twisted by the differential rotation between the disk and halo gases. The twist will amplify the field strength near the rotation axis, and will be observed as the galactic center jet (or tornado) of 4-kpc length.

5.8.2 Molecular Cylinder

Expanding ring features of various scales are found in the galactic plane, which are mainly observed in the molecular and hydrogen line emissions. The molecular expanding ring of 200 pc radius is the most typical example, which has a radius of 200 pc and is expanding at a velocity of 50 km s^{-1}. The tangential points to the 200-pc ring on the $l - v$ diagram appear at $l = 1°.6$ and at $l = -1°.2$. Cross-sections of the ring in CO emission at these longitudes indicate that they compose walls of 150 pc height perpendicular to the galactic plane with a thickness of about 20 pc. Namely, the expanding ring is a cylinder about 150 pc long. The kinetic energy of the expansion motion is on the order of 10^{54} ergs, which is comparable to that involved in the galactic center lobe. The coincidence between the energies in the 200-pc cylinder and the GCL is consistent if the GCL focuses on the ring in the future, and the 200-pc ring is the consequence of the shock focusing of a past GCL some 10^6 y ago.

A cylindrical molecular ring has also been found in the starburst galaxy M82. The "200-pc ring" in M82 is thought to be the consequence of an energy release, far more intense than that in our galaxy center, by successive explosions of supernovae in the central region of the dense molecular disk, which is followed by a vertical wind of high-temperature gas as well as the formation of an expanding ring and a cylinder. Existence of a vertical magnetic field is not known, whereas the intense nonthermal radio emission suggests field strength as high 10 μG.

5.9 Starburst and Nuclear Explosion

5.9.1 Bipolar Hypershells in Radio and X-Rays

The whole sky radio map at 408 MHz shows numerous radio spurs as shown in Fig. 5.17. The most prominent spur is called the North Polar Spur (NPS), which traces a giant loop in the sky of about 120° diameter, drawing a huge Ω over the galactic center. The brightest ridge of the shell at $l \sim 30°$ in the northern sky was recognized early in the 1950s, and was considered to be part of a nearby supernova remnant making a giant loop in the sky, called Loop I. Later the spur and loop were recognized as making the northern portion of the dumbbell, and the distance to the NPS is nowadays considered to be several kpc away, and the structures are related to the galactic center events.

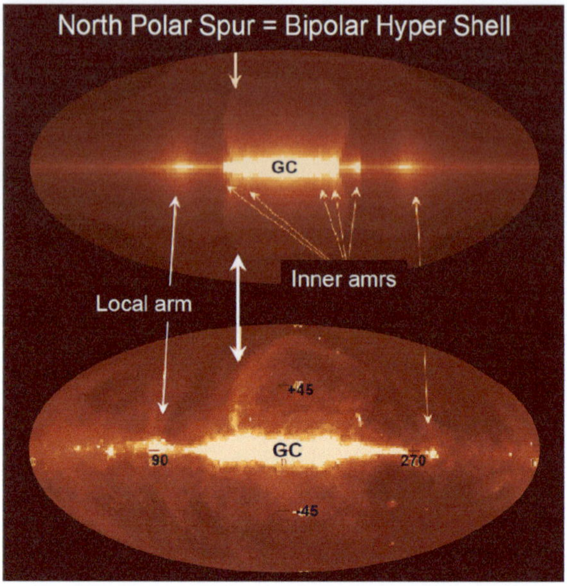

Fig. 5.17 The North Polar Spur in radio at 408 MHz (*Bottom*: Bonnj-Parkes All Sky Radio Continuum Survey) [3], compared with a bipolar-hypershell (BHS) model. A dumbbell-shaped shock front was produced by a starburst some 10^7 yr ago

The giant loop structure is also recognized in soft X-ray emission as observed by the ROSAT all-sky survey, revealing a bipolar shell structure in the sky (Fig. 5.18). The NPS and its counter spurs are well interpreted as due to a shock wave produced by the BHS model with total released energy on the order of $10^{55} \sim 10^{56}$ erg.

The galactic center is also associated with much higher-energy phenomena such as the Fermi bubbles [15], which are bipolar γ-ray emitting bubbles of diameter $\sim 50°$ emanating from the central region. The bubbles are located nearly concentric to the X-ray and radio loop of the NPS.

5.9.2 *Giant Shock Wave from the GC*

The prominent BHS feature in the sky, apparently emanating from the galactic center, is interpreted as due to an energetic explosive event(s) at the galactic center. A sudden energy release at the galactic center such as an explosion or a starburst results in vertical ejection of a gaseous structure, making huge shocked shells propagate in the galactic halo.

Figure 5.19 shows an example of numerical hydrodynamical simulations of the propagation of a shock wave produced at the galactic center. The shock wave expands spherically in the initial phase, but it expands more rapidly in the direction perpendicular to the disk plane because of the smaller gas pressure and mass in

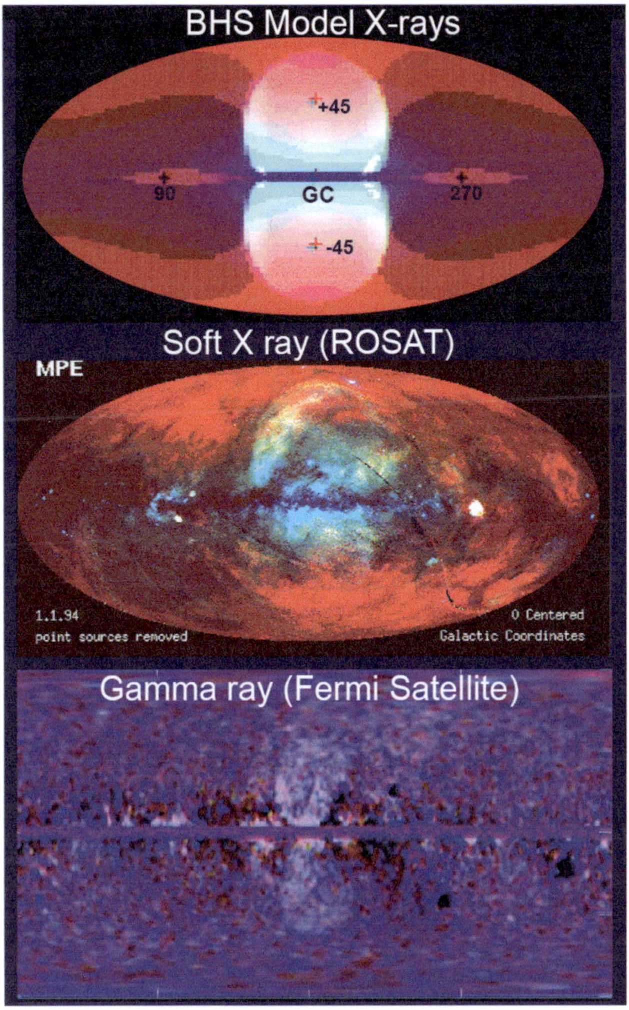

Fig. 5.18 Bipolar hypershell (BHS) in the galaxy (*top*), as compared with the ROSAT all-sky soft X-ray image [13] (*middle*), and the Fermi bubble in γ ray [15] (*bottom*)

this direction compared to the propagation in the disk plane. Then the shock front attains a bipolar Ω-shaped front elongated perpendicular to the disk producing a bipolar hypershell (BHS) structure. The shock form has high temperature and the high compression of magnetic fields, and radiates synchrotron radio emission and soft X-rays. Figure 5.20 shows the simulated result for soft-X ray emission and compares it with the ROSAT observation.

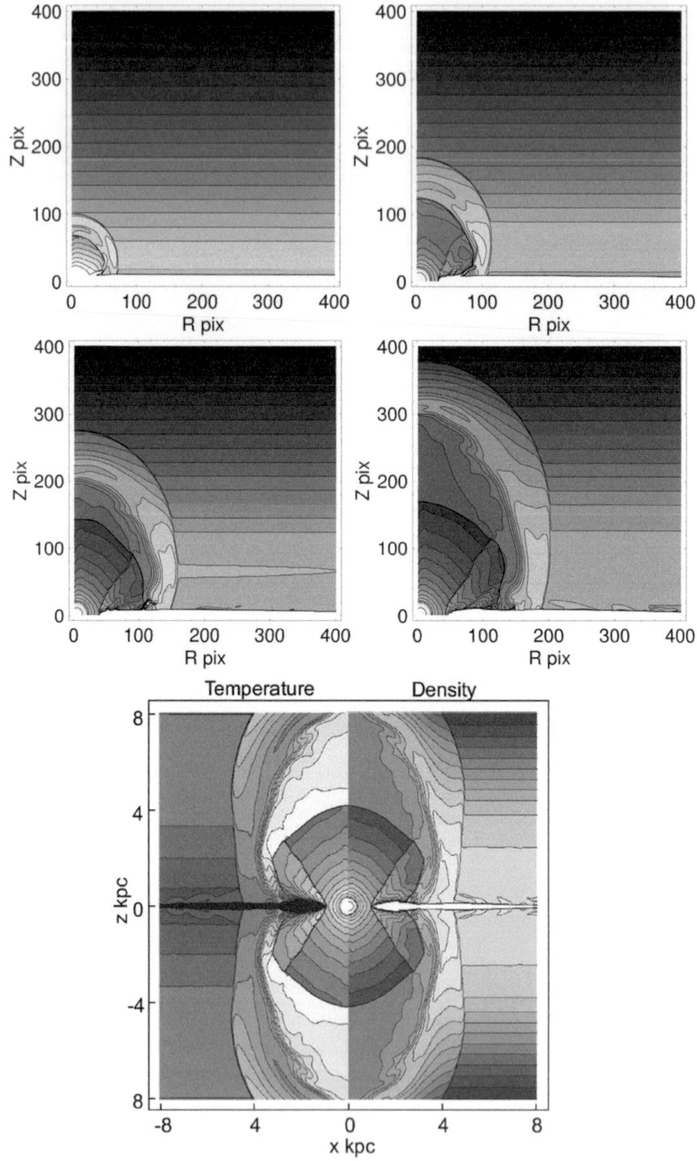

Fig. 5.19 Hydrodynamical simulation of evolution of density profile in bipolar hypershells (BHS) induced by starburst in the galactic center at elapsed time from the burst, $t = 2,\ 4,\ 6,\ 8 \times 10^6$ y, and density and temperature at 10×10^6 y. The shock wave produces huge dumbbell-shaped shells in the halo [14]. The compressed high- temperature shell radiates soft X-rays and radio continuum emissions, observed as the North Polar Spur

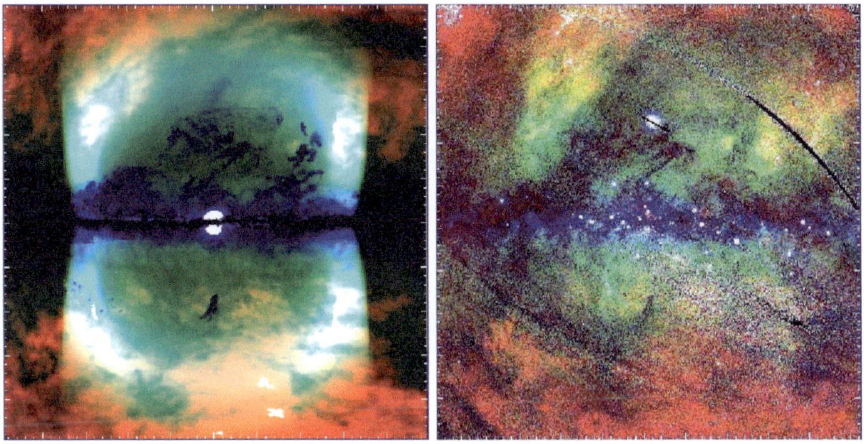

Fig. 5.20 Simulated X-ray image of the BHS compared with ROSAT observation. Intensities are color-coded by *red* for 0.25 keV, *green* for 0.8 keV, and *blue* for 1.5 keV

5.9.3 Definition of Starburst

By definition "starburst" is a phenomenon that a high rate of star formation (say $\sim 10^3$ times that in normal interstellar place) is realized in a specific volume of a galaxy within a much shorter duration (say $<10^8$ yr) than the timescale of the galaxy's evolution time. In this sense any star-forming regions in the galaxy that contain massive OB stars, HII regions, and molecular clouds may be regarded as starbursting regions when they are picked up individually. However, we here define a starburst galaxy as a galaxy that contains a coherent ensemble of star-forming regions within a region of ~ 1 kpc scale, in most cases near the galactic nucleus. In this section, we describe the properties of a starburst raising the typical starburst galaxy M82.

5.9.4 Appearance of a Starburst Galaxy

We summarize what the starbursting galaxy, and therefore star-forming regions, look like at various wavelength ranges (Fig. 5.21). Stars are born from dense molecular clouds that are seen most representatively in spectroscopic mm-wave radio observations of CO, HCN, H_2CO, HCO^+, OH, and so on. As the gas is to be supplied from the outer region of the galaxies by some dynamic mechanism, the galaxy must contain a rich envelope of gas often seen in the HI 21-cm line emission. Newly born stars, in particular OB stars, radiate intense UV and optical lights that ionize the very nearby gases to produce HII regions that emit thermal Bremsstrahlung (free-free) radio radiation with a flat spectrum. However, most of

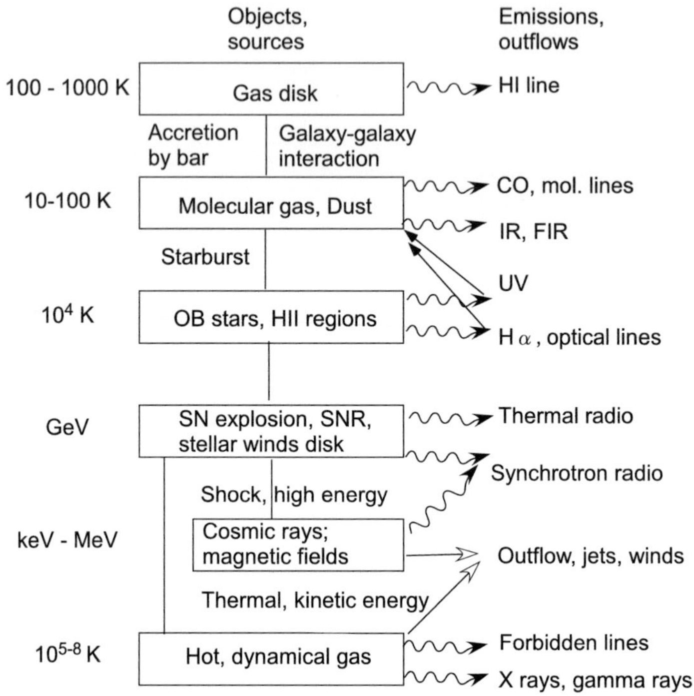

Fig. 5.21 Diagram illustrating what a starburst galaxy looks like

the UV and optical lights are absorbed by dust grains that are richly contained in the molecular clouds. The UV-heated dust of temperature 10–60 K re-emits black-body radiation in the far-infrared (FIR) domain peaking at $\sim$100 μm. Hence, the starburst region, and star-forming regions, are the strongest emitters of FIR radiation.

High-mass stars (OB stars) inject thermal and kinetic energies into the interstellar media through high-velocity stellar winds by their mass loss process. In addition, OB stars explode at the end of their lives as supernovae, and inject a great deal of energy into interstellar space through cosmic rays and shocked gas. Through the interaction with the interstellar magnetic field, which is in an energy equipartition with the turbulent gaseous component and cosmic rays, cosmic-ray electrons emit diffuse nonthermal (synchrotron) radio emissions with a steep spectrum. Remnants of supernova explosions are also strong nonthermal radio emitters, but are observed as individual compact sources. Finally, the heated interstellar gas radiates diffuse optical, UV, and X-ray emissions, depending on their temperatures. The thermal and kinetic energies, and therefore the pressure, often become large enough to blow off the disk gas, which is observed as an outflow of gas in the direction perpendicular to the galactic plane.

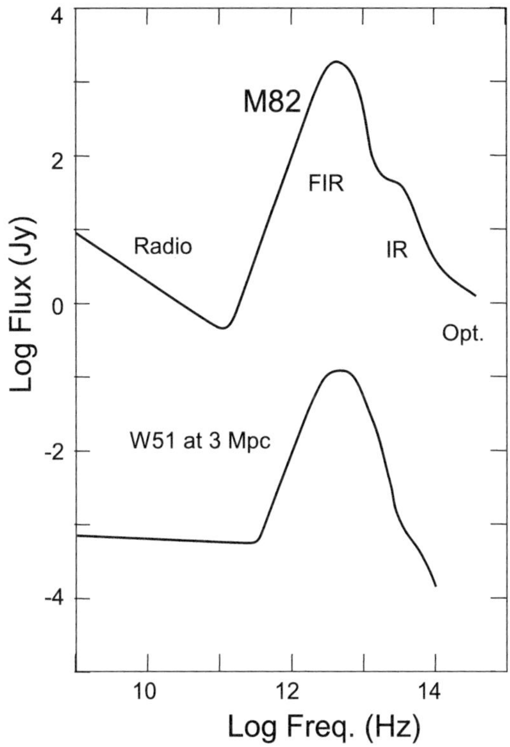

Fig. 5.22 FIR and radio spectrum of M82 compared with the galactic star-forming region W51 at the same distance of 3.25 Mpc

Figure 5.22 shows a comparison of spectra of typical starburst galaxy M82 and typical star-forming regions in our galaxy. As is readily seen the spectra resemble each other closely and without doubt the two sources are composed of similar, or almost the same, materials, except for the total luminosity. Namely, we need only a pack of $\sim 10^4$–10^5 objects equivalent to a galactic star-forming region such as W51 or W3 in the central ~ 1 kpc region in order to reproduce the emissions of M82 from radio to infrared.

In Table 5.2, we compare the FIR luminosities of the central 1 kpc of M82 with that of the central 1 kpc of our galaxy. We also compare their total masses and molecular gaseous masses. Obviously M82 radiates $\sim 10^2$ times more energetic radiation in the FIR domain. As the FIR emission is an almost perfect indicator of energy released by OB stars, we may conclude that high-mass stars are produced there about a hundred times more than in the same volume of our galactic center region, although the gas contents are almost the same. This means that the efficiency of star formation per unit of gas mass is $\sim 10^2$ times higher in M82. We also note that the fraction of mass in the form of gas in M82 is much higher than in the galaxy.

Table 5.2 Comparison of the ISM in the central 1 kpc of M82 and the Milky Way [5], p200, 201

	M82	Galaxy
FIR luminosity, $L_{\rm FIR}$	$4 \times 10^{10}\,L_\odot$	$5 \times 10^{8}\,L_\odot$
Dynamical (total) mass, $M_{\rm dyn}$	$10^{9}\,M_\odot$	$10^{10}\,M_\odot$
Molecular hydrogen mass, $M_{\rm H_2}$	$2 \times 10^{8}\,M_\odot$	$2 \times 10^{8}\,M_\odot$
$L_{\rm FIR}/M_{\rm tot}$	$40\,L_\odot/M_\odot$	$0.05\,L_\odot/M_\odot$
$M_{\rm gas}/M_{\rm tot}$	0.2	0.02

5.9.5 Implication of Starbursts

Starburst galaxies, for their high luminosities at any frequency, provide the opportunity to detect and measure the emissions more easily compared to normal galaxies, and are much easier to be found, even if they are at cosmological distances. As is often the case, the starburst galaxies exhibit peculiar morphological features such as ejection of matter or some explosive nature, and show many dynamic characteristics. Many observers are attracted by such enthusiastic properties. Because of the high luminosity and therefore high signal-to-noise ratio in observations, as well as by attracting many observers to the challenge, a wealth of good quality data at various wavelengths is available, in particular for such nearby starburst galaxies as M82 and NGC 253. This provides us a good opportunity to investigate their physical properties in great detail.

The physics that must be studied, of course, is the mechanism and origin of the active star formation. To do this, we need to know such parameters as the efficiency and rate of star formation from molecular clouds, initial mass function of stars being born there, the lifetime of the bursting phase, and evolution. To determine these quantities by measuring the emissions from bursting regions, we must first see the physical conditions in the regions, namely the temperature, excitation condition, optical depth of the emission, density and size of gas clouds, and the like; also the transfer of emission is an important subject.

As a bursting galaxy is in a violent phase, it is essential to understand its dynamic circumstances, namely the motions of gas, the existence of shock waves, the outer inflowing motion, the turbulence, and so on. Information of such dynamic phenomena can be obtained only through spectroscopic observations of radio line emissions such as HI and CO as well as from optical line emissions such as the Hα and forbidden lines.

A further implication of the study of starburst galaxies is their role in the evolution of normal galaxies. Namely, whether a starburst is really a peculiar phenomenon occurring in some very specific galaxies, or if it occurs rather commonly in any galaxy during its evolution. Does a starburst occur only once during the galaxy evolution, or is it a recurrent phenomenon? All these questions may significantly affect understanding the chemical evolution of galaxies, for the drastic increase in the star-formation rate causes a significant change of the interstellar chemical abundance on which the evolution of a galaxy significantly depends.

Fig. 5.23 The starburst galaxy M82 with Hα jets perpendicular to the galactic plane (APOD 2016 http://apod.nasa.gov/apod/ap160221.html; NASA, ESA, and The Hubble Heritage Team, STScI/AURA)

5.9.6 *Starburst History in M82*

M82 (NGC 3034; Arp 337; 3C231) was recognized as a peculiar edge-on galaxy in the M81 group of galaxies early in the 1950s and was known to be a companion to M81 (Fig. 5.23). M82 was classified as Irr II. From its peculiar morphology exhibiting numerous filamentary structures in emission and in absorption extending out of the galactic disk, the galaxy was suggested to be either exploding or colliding. In the 1960s an explosion hypothesis was advocated by optical measurements of the velocity gradient along the minor axis in the Hα emission. A high-expansion velocity of $\sim$500 km s^{-1} was reported that suggested a gigantic explosion some 10^6 yr ago at the nucleus.

In the 1970s, linear polarization measurements of scattered light in the halo showed the existence of a wealth of dust high above the galactic plane over 2 kpc. However, the detection of red-shifted scattered light of the Hα line brought some complications in the interpretation of the expanding-halo model on the basis of a moving mirror hypothesis.

A large-scale extension of gas around the galaxy was also known from the HI line observations. The HI velocity showed a large-scale distortion from the circular motion of the gas in the HI halo, strongly suggesting a tidal interaction with the galaxy M81, and hinted at a triggering dynamic mechanism of the bursting star formation in the inner region. In fact, a prominent HI bridge connecting M81 and M82 has been observed. More inner regions were mapped in HI absorption against the radio continuum emission showing a dense HI condensation in the central 2×0.3 kpc region.

In the 1980s, ground-based and satellite observations at infrared wave range became available. Near- and mid-infrared observations (2–20 μm) have shown that IR characteristics of M82 and NGC 253 are well understood by a starburst hypothesis. FIR observations became available and thorough IR spectra of these galaxies became known, addressing the similarity between galactic star-forming regions and the galaxies in the IR domain. The starburst hypothesis could well explain the wealth of young stellar clusters detected in optical emission line observations. Of course, the IRAS observations put M82 among the most luminous FIR sources in the sky. The IRAS discovery of luminous FIR galaxies, far more luminous than M82 in some cases, forwarded much in ascertaining the importance of the study of starburst galaxies, because most of them could be of similar nature to M82 as is readily indicated from their similar FIR spectra.

X-ray imaging of M82 revealed that a high-temperature ($\sim 10^7$ K) gas coexists in the central region with the hot gas extending perpendicularly to the galactic plane toward the halo as high as $\sim$1 kpc. This suggested an energetic heating mechanism of interstellar gas in the nuclear region. Indeed, high-resolution radio-continuum mapping of the galaxy has shown a large number of young supernova remnants of $\sim 10^2$ yr or rather less concentrating in the central 400×100 pc region. This suggests a high rate of supernova explosions, $\sim$one SN per a few years, hinting at the energy source of the high-temperature component. Earlier continuum mapping showed an extended nonthermal emission, which indicated a magnetic field as strong as 35 μG mixed up with high-energy cosmic-ray particles.

Being stimulated by those observation results and by the high-resolution CO observations described later, several theoretical models have been proposed to simulate the large-scale outflow of gas and wind driven by the starburst activity in the center.

5.9.7 CO Observations of M82

M82 contains a large amount of molecular gas and was the first detected galaxy in the molecular lines (Tables 5.3 and 5.4). In the early 1980s, detailed CO line observations showed that the molecular gas in the central region of M82 is optically thin for the CO line, suggesting anomalously high- velocity dispersion. Wider CO mapping also showed a large-scale molecular halo and a velocity gradient along the minor axis, suggesting a significant noncircular motion and/or outflow.

Large-aperture single dishes, for example, the NRO 45-m telescope and IRAM 30-m telescope, and mm interferometers have made it possible to obtain high-resolution maps with resolutions of 5–15$''$ (Fig. 5.24), and provided the opportunity for more detailed comparisons with other observations at various wave ranges.

Table 5.3 Summarized masses of the various contents in M82

Material	Mass	Area
Mol. gas (H_2)	$1.1\times10^8\,M_\odot$	at $r < 700$ pc
Atomic gas HI	5×10^7	$r < 700$ pc
Ionized gas (HII)	7×10^7	$r < 600$ pc
Dust	5×10^5	$r < 300$ pc
Dyna. mass	10^9	$r < 700$ pc

Table 5.4 The observed parameters for the outflowing molecular gas

Mass (H_2)[a]	$5\times10^7\,M_\odot$
Velocity[b]	$\sim$300 km s^{-1}
Height	$> \pm$ 500 pc
Kinetic energy	$\sim5\times10^{5}5$ erg

[a]Within the region of $|z| > 100$ pc and $|r| < 500$ pc
[b]Assuming the inclination angle of 82° (90° is edge-on)

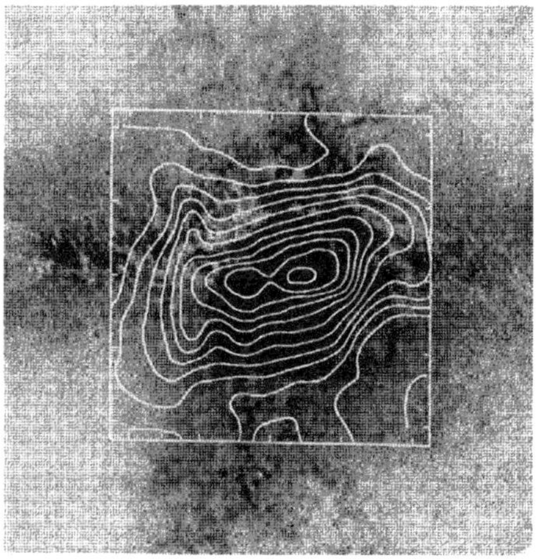

Fig. 5.24 CO intensity map of M82 taken with the NRO 45-m telescope [8]. Inset is a Hα photograph

Figure 5.25 shows the integrated intensity map of the ^{12}CO ($J = 2-1$) line emission of M82 compared with maps at various wavelengths. The distribution of the emission is almost the same as that obtained in the $J = 1-0$ line, and most of the structures seen in the 45-m CO map are seen clearly in this figure, for example, the two peaks and the spurlike extensions toward the halo.

Comparison of ^{12}CO ($J = 2-1$) line profiles with those in $J = 1-0$ transition profiles for the 200-pc ring region and for outer regions shows that the inner region is optically thin for the CO ($J = 1-0$) line with an optical depth of $\tau \sim 0.5$ or assumed same excitation temperature of $T_{exc} = 40$ K as the dust. On the other hand, in the outer regions, the emission is optically thick.

Fig. 5.25 Comparison of maps of M82 in various emissions [8]

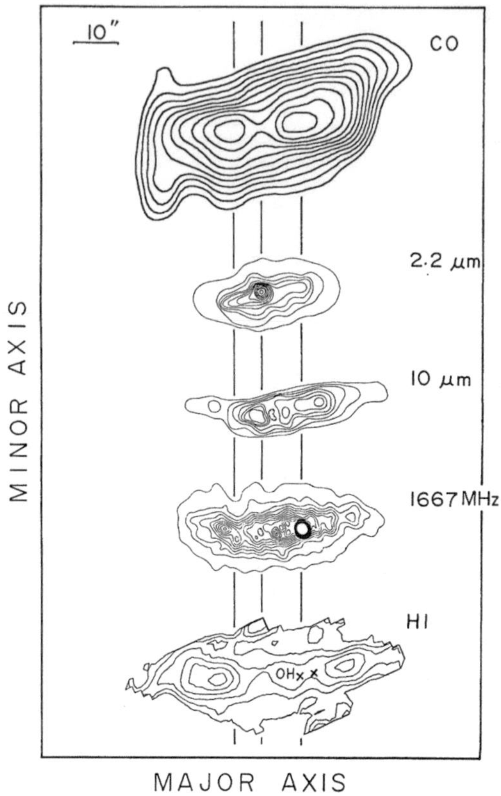

5.9.8 *Comparison with Other Observations*

The CO observations have made it possible to compare in detail the molecular hydrogen distribution with other observations at various wavelengths from X-ray to radio as shown in Fig. 5.25.

5.9.8.1 IR and FIR Emissions

The 2.2 μm map shows the distribution of low-temperature stars such as supergiants associated with circumstellar dust shells of high temperature, 500–800 K. The nucleus represented by the 2.2 μm peak is located in the minimum position of the CO emission between the two peaks. This shows that the central star cluster is located in the center of the ring where the molecular gas density is much lower, whereas the extended 2.2 μm emission feature coincides with the molecular ring and shows the existence of the stellar population mixed up with the molecular gas clouds there.

The 10 μm emission shows a similar distribution to that of the molecular hydrogen. The 10 μm emission comes from dust of temperature $\sim$100 K mixed with ionized HII gas near star-formation regions heated by O and early B stars. In the present resolution this hot dust component coexists with the molecular ring composed of warm (40 K), small clouds. The warm dust component radiates in the FIR range peaking around 100 μm. The energy source of the FIR emission is the absorbed UV and optical light from OB stars.

5.9.8.2 HI Gas

HI gas distribution as determined from the 21-cm absorption features also has two peaks that coincide with the CO peaks as determined from the interferometer observations, showing that the HI gas also composes a ring. The total mass of HI gas, and therefore the density, involved in the central 1 kpc is about half that of molecular hydrogen. Such high density of HI, comparable to H_2 density, in the central region seems extraordinary, because normal galaxies generally have a deep depression (or a hole) of HI in their centers and H_2 (molecular gas) dominates there. The high HI content in the central ($<$500 pc) region of M82 may suggest an intense inflow (accretion) of gas from the outer region of the HI envelope.

5.9.8.3 Radio Continuum Emission

Radio continuum emission of M82 consists of two components. The extended component is mainly due to synchrotron radiation by the strong interstellar magnetic field ($\sim$ 35 μG) and cosmic rays and partly, in particular in the mm wave range, because of the thermal (free-free) emission from ionized gas. Another component consists of numerous discrete sources of small diameter, which are mostly attributed to young supernova remnants. The global distribution of the two components is essentially the same and coincides well with the CO intensity distributions. However, the two peaks observed in the CO line are not clearly seen in the continuum emission, which is more uniformly distributed.

5.9.8.4 Optical Images

In Fig. 5.25 the CO intensity distribution is compared with the optical appearance of M81. It is interesting to compare the spurlike CO features in the halo, presumably a molecular cylinder, with the optical filamentary structures. The running optical filaments are located within the wall of the molecular cylinder. The optical filaments are a mixture of Hα emission from the high- velocity ionized gas inside the cylinder and dust filaments associated with the molecular cylinder that surrounds the high-temperature flow.

5.9.8.5 X-Ray Emission

The X-ray emission is elongated along the minor axis, originating near the center of the molecular 200-pc ring. The CO cylinder surrounds the X-ray feature. The X-ray emission indicates that a high-temperature gaseous component is produced in the central region, heated by energetic phenomena associated with SN explosions during a starburst. The configuration elongated perpendicular to the galaxy plane suggests that the gas is outflowing toward the halo, collimated by the molecular wall.

5.9.9 NGC 253

NGC 253 is the second most active starbursting galaxy among the nearby galaxies exhibiting high IR luminosity. CO observations at a resolution $\sim 30''$ have shown a high concentration of molecular gas toward the center and a molecular gas ring of radius 3 kpc. However, no higher resolution observations have been made therefore no detailed distribution of molecular gas in the nuclear disk is known as yet, partly because of the observational difficulty due to its low declination.

Measurements of the velocity field in the central region have shown a large noncircular motion indicating a barred structure. Evidence for outflow of gas from the nuclear region has been obtained by optical and radio emission line measurements. X-ray observations show an extended emission associated with the nuclear region elongated in the direction perpendicular to the major axis, indicating a vertical outflow of high-temperature gas, similar to the M82 case. All these observations, in particular the high IR luminosity, show that the nucleus is active and a starburst is taking place.

Recently a large-scale vertical outflow of gas was suggested by a detailed inspection of dust filaments in optical photographs (Fig. 5.26). Numerous dark filaments of length more than 1 kpc are running vertically to the major axis in the inner region of radius 2–3 kpc. The vertical filaments are either caused by a vertical magnetic field accelerating cold gas from the disk toward the halo, or caused by a vertical outflow (or wind) driven by starburst activity in the central 1–2 kpc.

5.10 Starburst Models

5.10.1 Ring-and-Outflowing Cylinder Model

The results of the various observations can be interpreted in terms of a ring-and-outflowing cylinder model, and are schematically illustrated in Fig. 5.27. The molecular hydrogen gas in the central region of M82 consists of two major components; the “200-pc ring” in the galactic plane and the cylindrical halo with high-velocity outflow.

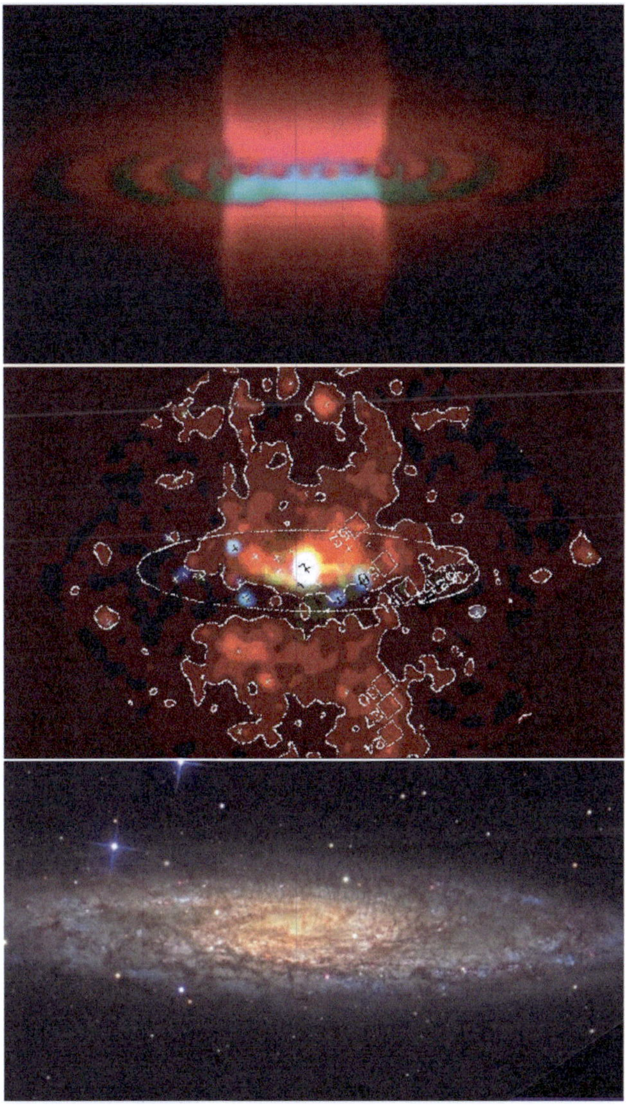

Fig. 5.26 Starburst galaxy NGC 253. (*Top*) Simulated X-ray BHS; (*Middle*) ROSAT X-ray image [11]; (*Bottom*) Optical image, showing vertical dusty filaments that indicate outflows from the disk (APOD 2006, http://apod.nasa.gov/apod/ap060421.html, Credit & Copyright: R. Jay GaBany)

The 200-pc ring is associated with spurlike features that can be deconvolved to a cylindrical distribution of molecular gas in an axisymmetric case. The total molecular mass involved in the cylinder is about $5 \times 10^7 M_\odot$ which is the mass integrated above 100 pc of the galactic plane. The radius of the cylinder is about 300 pc with thickness ~200 pc. It extends perpendicular to the disk plane as high as

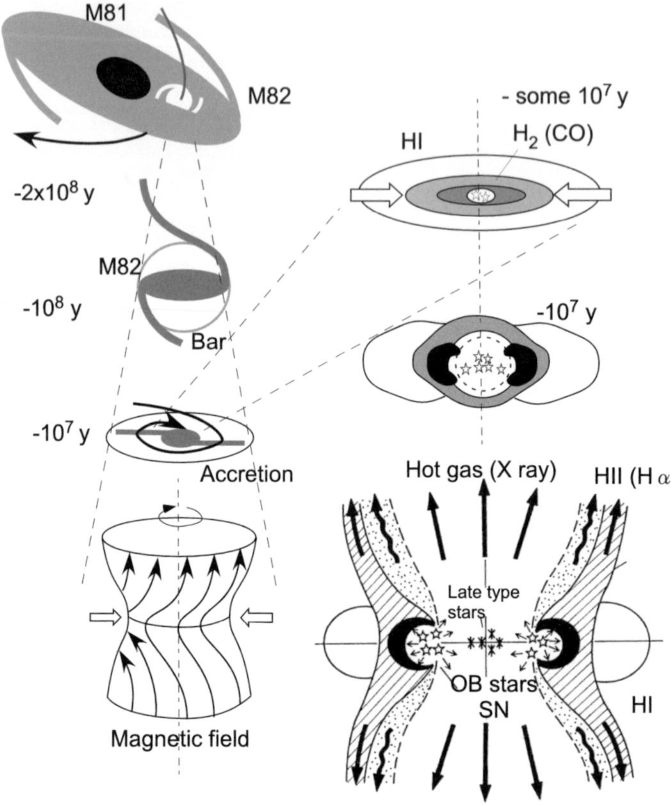

Fig. 5.27 Evolution of M82 after tidal encounter with M81, and accretion of gas into the center to cause a starburst, forming a gaseous ring and cylinder structure

$\sim$1 kpc. From the velocity gradient perpendicular to the disk plane, the gas in the cylinder is flowing out of the plane at a velocity $\sim$300 km s^{-1} for the inclination of $i = 82°$. The total kinetic energy of the outflow is about 10^{56} erg.

The energy of the expanding motion can be supplied by SN explosions. The SN explosion rate in the ring is estimated from radio continuum luminosity, which is about one SN per a few years. The energy given to the interstellar gas is estimated by the SN rate as as

$$E_{\rm kin} = \eta E_{\rm SN} R_{\rm SN} t_{\rm cyl}, \tag{5.1}$$

where *eta* is the efficiency of energy transfer to the gas, $E_{\rm SN} \sim 10^{51}$ erg is the energy of one SN explosion, $R_{\rm SN} \sim$ 1–0.3 yr^{-1} is the SN rate and $t_{\rm cyl} \sim Z/V_{\rm exp} \sim$ 1 kpc / 300 km s^{-1} $\sim 3 \times 10^6$ yr. The conversion rate, η, may be taken as $\sim$0.1. Then we obtain $E_{\rm kin} \sim 10^{56}$ erg, comparable to the observed value of outflowing kinetic energy.

It is likely that the gas clouds are stretched in the direction of the flow. In fact, optical photographs show numerous filamentary dust lanes being stretched perpendicular to the disk, which are most likely caused by such stretched molecular gas clouds in the cylinder by the high-velocity flow. The observed peculiar morphology of the vertical Hα filaments is explained by the cylinder model.

At the contact surface of the low-temperature molecular wall and high- temperature ionized gas (X-ray halo) there appears a transient region with temperature 10^{4-5} K, which radiates Hα emission. The Hα emission is seen through the molecular wall containing numerous stretched dust clouds silhouetted against the Hα emission. The mixture of Hα and dust filaments are thus observed as running perpendicular to the galactic plane.

Thus from the viewpoint of energetics the observed high-velocity molecular outflow perpendicular to the disk and the extended X-ray emission can be naturally attributed to the intense energy release by the starburst in the nuclear ring. The peculiar optical morphology with Hα and dust filaments can also be explained by this cylinder outflow model. In the next section we consider this model in a more quantitative way on the basis of numerical simulations.

5.10.2 Bipolar Hypershell and Super Wind Models

The outflow phenomena of gas in M81 and NGC 253, which are typical starburst galaxies, can be modeled by various mechanisms related to the formation of extragalactic jets.

The simplest model is the bipolar hypershell and/or wind model. In this model, a strong shock wave, produced by accumulating SN explosion fronts, encounters the ambient interstellar gas and halo. The shock propagates much faster in the direction perpendicular to the disk plane, blowing off into the halo, resulting in a collimated flow along the minor axis, while it propagates slowly in the disk blocked by the dense gas in the disk.

Accordingly, the shock wave propagates faster in the direction perpendicular to the disk and the disk is blown off toward the halo, forming an elongated Ω-shaped shock front in the halo, partly composing a cylindrical structure (Fig. 5.19).

The shock wave model in a disk and halo system can already explain some essential features observed in M82: the ring structure in the disk plane and the cylindrical collimated flow perpendicular to it. The high temperature in the cavity surrounded by the shock front may explain the X-ray emission. For the cold molecular disk to form, the cooling time in the shock-compressed region must be short enough compared with the timescale of the flow. In the case of M82 the cooling time must be shorter than $\sim 10^7$ yr, which is realized even in the halo region where the free-free cooling time is estimated to be $\sim 10^6$ yr.

5.10.3 *Magnetic Model*

Radio continuum observations showed the existence of strong magnetic fields of ~35 μG, although the observations do not tell about the orientation of the magnetic field. It was shown that the central region of a disk galaxy likely possesses a strong vertical magnetic field running perpendicular to the disk plane, often dominating over the field parallel to the galactic plane. By such a strong vertical magnetic field the flow will be better collimated in the direction parallel to the field.

A magnetically driven outflow model has shown that an accretion of gas in the disk twists the vertical magnetic field. The twist propagates toward the halo along the field, accelerating the disk gas perpendicular to the disk, forming a twisted magnetic jet. The magnetic model needs no violent heating of gas but can create even a cold (molecular) gas flow in a vertical cylinder.

This may better explain the molecular gas cylinder in M82, although it is hard to account for the central high-temperature gas, for the model does not require any explosion or strong energy release at the center.

In reality, however, the wind/explosion model and the magnetic twist model may be combined. The intensive accretion of gas toward the center, which is necessary to trigger the starburst (see the next section), would produce a magnetic-twist acceleration of cold gas from the dense disk, creating the molecular gas cylinder. The starburst causes a high-temperature gas flow in the inner region.

5.10.3.1 Bar and Starburst Trigger

The tidal encounter of galaxies produces a barred structure in the innermost region of the disturbed galaxy (Fig. 5.28). In the case of M82 the following scenario of the formation of bar and starburst triggering has been proposed. In some 10^7 yr after the per-galactic passage of M81, a bar structure was produced in the M82 disk, and the bar lasted more than several rotations ($>10^8$ yr). By the noncircular shocked motion of gas, a rapid inflow was enhanced toward the central region, where the accreted gas made a dense gaseous core around the nucleus.

The star-formation rate (SFR) in a galaxy has been shown to be related to the gas density through Schmidt's law

$$SFR \sim \rho^a \,. \tag{5.2}$$

From statistics of OB associations, HII regions, and blue light distributions in comparison with the HI and molecular gas distributions, the power a has been estimated as 1~2. If cloud-cloud collisions are taken into account, in particular, in such a highly turbulent region as the nuclear disk of M82, a modified Schmidt's law may be necessary, written as

$$SFR \sim \rho^a \sigma_v^b, \tag{5.3}$$

Fig. 5.28 (**a**) Simulation of tidal encounter by another galaxy, which causes a self-sustaining stellar bar [9]. Gas is accreted to the center by angular momentum loss due to bar-induced shocks. Time is measured in rotation period at the tidal radius from the peri-galactic passage. (**b**) Detailed simulation of bar-shocked gas flow and accretion [12]

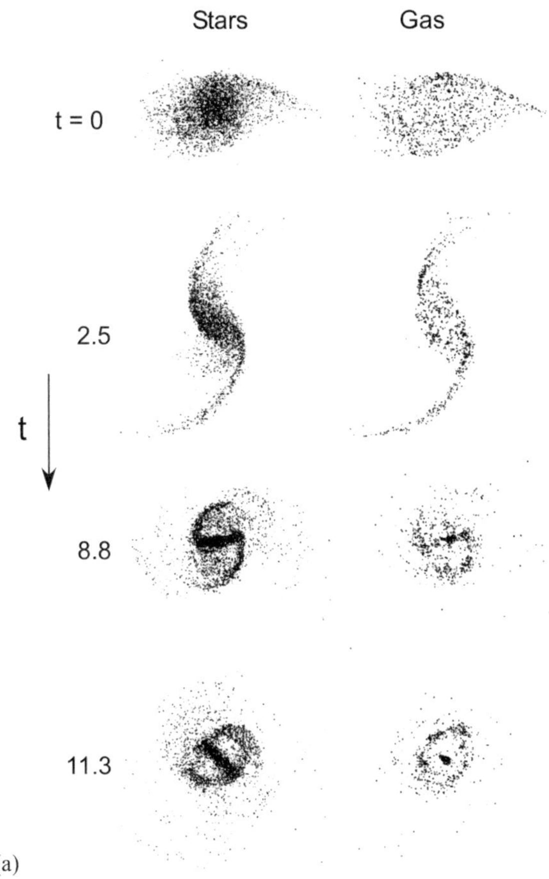

(a)

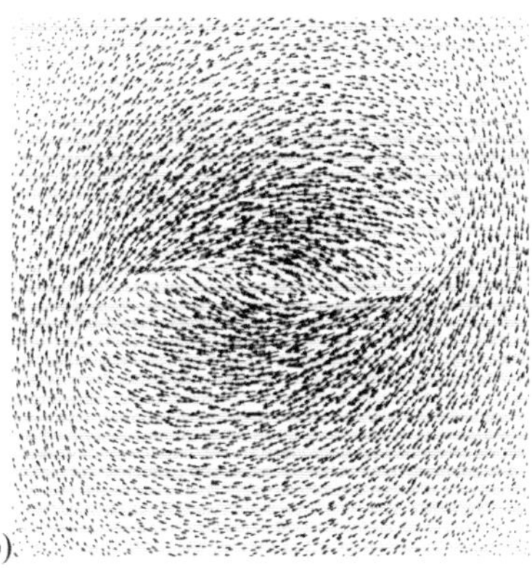

(b)

where σ_v is the velocity dispersion that acts to increase the collision frequency among clouds.

If cloud collision is the major factor, the power b is about unity. If the shock compression in a cloud caused by collision, which is proportional to the square of the Mach number for an isothermal shock, is taken into account, the power b will be higher. We may take b as between 1 and 3.

In the central region of M82 the average gas density is $\rho \sim 10\text{–}100\ \mathrm{H\,cm^{-3}}$ and the velocity dispersion $\sigma_v \sim 50\text{–}100\ \mathrm{km\,s^{-1}}$, which are both 5 to 10 times those in normal interstellar space in galaxies. For $a \sim 2$, $b \sim 2$, we obtain SFR about 100 to 1000 times that in normal galaxy disks. Such a high SFR is attained within a few 10^7 yr triggered by the sudden accretion of gas by a bar.

The high SFR results in a high rate of birth of massive stars (OB stars) which radiate intense UV and optical lights. The lights are absorbed by the dust in molecular clouds, heat them, and are reradiated in the IR emission. The high rate of SN explosions and intense stellar winds from OB stars follow. The energy thus released and injected into the interstellar medium is then transformed to thermal and turbulent (kinetic) energies of the gas, creating high-temperature ionized gas and large dispersion of the clouds. According to the heating of the gas, the pressure increases, which results in compression of gas in the 200-pc ring and in an outflow of gas from the disk.

The shock compression results in increases both in the density and collisions. This may further increase the SFR, and more extensive star formation is triggered. This chain reaction of star formation, or the starburst, will last until the gas clouds in the nuclear disk are all exhausted or blown off from the disk through the galactic wind. In this sense the starburst (chain-reacting star formation) is an ensemble of numerous star-forming regions where the sequential star formation is proceeding.

5.10.4 Unified Scenario of Starburst

The starburst process may be summarized as follows, and is illustrated in the diagram in Fig. 5.29.

(i) Galaxy-galaxy collision (encounter) causes a strong tidal disturbance in the companion galaxy. The disturbance is stronger when the disturber is massive and the orbit is "direct" (orbiting in the same sense as the rotation of the disturbed).

(ii) The tidal disturbance produces a long-standing (self-sustaining) barred structure in the innermost region of the disturbed galaxy.

(iii) The barred potential causes a strong galactic shock in the rotating gas, resulting in a rapid accretion into the nuclear region. The accretion creates a dense core in the center of the disk. Coupled with a strong vertical magnetic field, the accretion triggers a vertical outflow of cold matter from the disk through the twist of the magnetic lines of force.

Fig. 5.29 Schematic diagram summarizing a possible starburst mechanism

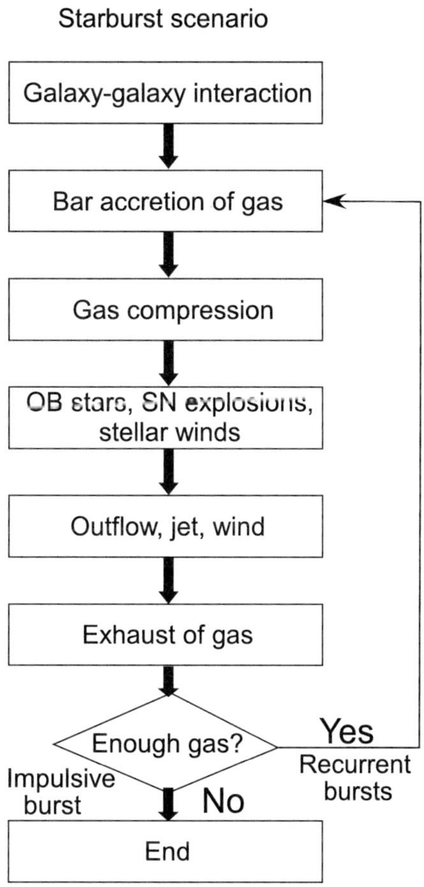

(iv) The increase in the gas density and turbulent motion in the central region results in a sudden increase of SFR according to the modified Schmidt law, and new stars are intensively born.

(v) UV and optical lights from thus-born massive stars are absorbed by dust clouds, heat them, and are reradiated in intense IR emission.

(vi) Stellar winds from massive stars and SN explosions excite the interstellar medium to a high-temperature and highly turbulent state. This results in a rapid expansion of the central region and forms a shock-compressed ring.

(vii) According to the compression in the ring the SFR increases and further star formation is enhanced. Stages (iv) to (vii) are repeated in a chain-reacting way.

(viii) When the pressure of heated gas becomes large enough, the upper part of the disk is broken and blown off vertically, and a collimated flow of high-temperature gas is formed. The flow is surrounded by the cold (molecular) gas cylinder that was partly accelerated via the magnetic twist by the accretion.

(ix) Through the outflowing wind and expansion of the ring in the disk, the central gas is exhausted. As the overall gas density decreases, the SFR decreases and the starburst stops.

(x) If the galaxy contains no more gas in the surroundings or no more trigger (stage (i) through (iii)) is expected as in the hyperbolic orbit case of a galaxy encounter, the starburst finishes all at once. This type may be called an impulsive starburst (ISB).

(xi) If there exists enough gas in the surroundings and if either the galaxy encounter repeats on an elliptical orbit or the bar structure is self-sustaining for a longer time, the stages from (i) or (ii) will repeat. This type may be called a recurrent starburst.

(xii) A shorter-period starburst cycle is also possible, when the gravitational potential of the central region is large so that the gas blown off by a starburst falls down again, as suggested for our galactic center region. This is another type of recurrent SB.

5.10.4.1 Bar and Starburst in the Large Magellanic Cloud

The Large Magellanic Cloud (LMC) comprises a close dynamical system with the SMC (Small Magellanic Cloud) and the galaxy, which are gravitationally interacting with each other. It was shown that LMC and SMC experienced a close collision some 10^8 yr ago.

It is therefore likely that the tidal interaction caused a bar in the central region of LMC, which is indeed observed as a stellar bar located off-center from the kinematical center of HI gas.

It was also suggested that a burst of star formation occurred some 10^8 yr ago triggered by the interaction, which is indeed confirmed by the detection of many blue globular clusters. Also, the LMC is known to have some peculiar activity of star formation even today, as observed as the 30 Dor complex.

This burstlike activity may have resulted in accretion of gas from the outer region and in the anomalous compression of gas around 30 Dor. In fact, we note that LMC and SMC are gas rich, particularly in HI, and are embedded in an extended HI envelope.

References

1. Genzel, R., Hollenbach, D., Townes, C.H.: The nucleus of our Galaxy. RPPh **57**, 417 (1994) [BH in GC]
2. Gillessen, S., Eisenhauer, F., Trippe, S., et al: The orbit of the star S2 around SGR A* from very large telescope and keck data. ApJ **692**, 1075 (2009) [BH in GC]
3. Haslam, C.G.T., Salter, C.J., Stoffel, H., Wilson, W.E.: A 408 MHz all-sky continuum survey. II – the atlas of contour maps. AA Suppl. **47**, 1 (1982) [408 MHz All Sky Radio]

4. Lo, K.Y., Claussen, M.J.: High-resolution observations of ionized gas in central 3 parsecs of the Galaxy – possible evidence for infall. Nature **306**, 647 (1983)
5. Lugten, J.B., Watson, D.M., Crawford, M.K., Genzel, R.: The interstellar medium in the central 1 kiloparsec of M82. ApJL **311**, 51 (1986)
6. Morris, M., Yusef-Zadeh, F.: Unusual threads of radio emission near the galactic center. AJ **90**, 2511 (1985) [GC magnetic field]
7. Miyoshi, M., Moran, J., Herrnstein, J., et al.: Evidence for a black hole from high rotation velocities in a sub-parsec region of NGC4258. Nature **373**, 127 (1995) [Black hole in NGC 4258]
8. Nakai, N., Hayashi, M., Handa, T., Sofue, Y., Hasegawa, T., Sasaki, M.: A CO absorption line toward M82 – a large molecular halo or intergalactic molecular gas? PASJ **38**, 185 (1986) [M82 CO]
9. Noguchi, M.: Gas dynamics in interacting disc galaxies – fuelling of nuclei by induced bars. AA **203**, 259 (1988) [Bar formation by interaction]
10. Oka, T., Hasegawa, T., Sato, F., Tsuboi, M., Miyazaki, A.: A large-scale CO survey of the galactic center. ApJS **118**, 455 (1998)
11. Pietsch, W., Vogler, A., Klein, U., Zinnecker, H.: X-ray observations of the starburst galaxy NGC 253 – II. Extended emission from hot gas in the nuclear area, disk, and halo. AA **360**, 24 (2000) [NGC 253 ROSAT]
12. Roberts, W.W., Jr., Huntley, J.M., van Albada, G.D.: Gas dynamics in barred spirals – gaseous density waves and galactic shocks. ApJ **233**, 67 (1979) [Bar shocked inflow]
13. Snowden, S.L., et al.: ROSAT survey diffuse x-ray background maps. II. ApJ **485**, 125 (1997) [ROSAT All Sky X-rays]
14. Sofue, Y.; Habe, A.; Kataoka, J. et al.: Galactic Centre hypershell model for the North Polar Spurs. MNRAS **459**, 108S (2016) [GC explosion and BHS]
15. Su, M., Slatyer, T.R., Finkbeiner, D.P.: Giant gamma-ray bubbles from Fermi-LAT: active galactic nucleus activity or bipolar galactic wind?. ApJ **724**, 1044 (2010) [Fermi Bubble]
16. Tsuboi, T., et al.: Prominent polarized plumes in the galactic center region and their magnetic field. AJ **92**, 818 (1986) [GC radio polarization]
17. Yusef-Zadeh, F., Morris, M.: The linear filaments of the radio arc near the Galactic center. AJ **94**, 1128 (1987) [VLA Radio continuum]

Chapter 6
Nonthermal Emission and Magnetic Fields

6.1 Synchrotron Emission and Linear Polarization

6.1.1 Synchrotron Intensity and Magnetic Field Strength

The intensity of synchrotron (nonthermal) emission is related to the electron density N, electron energy E, and the magnetic field strength B as

$$I_\nu \propto B^2 E_E^2 N. \tag{6.1}$$

On the assumption that the magnetic and cosmic-ray energy densities are in equilibrium,

$$\frac{B^2}{8\pi} \sim N_E E \left(= \int_{E_1}^{E_2} E\,N(E) dE \right) \tag{6.2}$$

and

$$\nu \propto BE^2, \tag{6.3}$$

we have an approximate formula that relates B and intensity I of the emission:

$$B \sim 3.3 \times 10^2 \nu_{1\,\mathrm{GHz}}^{-1/7} \epsilon^{2/7} \ \ (\mathrm{G}), \tag{6.4}$$

where

$$\epsilon = \frac{\int I_\nu d\nu}{L} \sim \frac{\nu I_\nu}{L}\,(\mathrm{erg\,s^{-1}\,cm^{-3}}) \tag{6.5}$$

The original version of this chapter was published without printing chapter abstract. The abstracts for each chapter are updated in the front matter in the current version of the book. An erratum to this chapter can be found at https://doi.org/10.1007/978-981-10-3445-9_7.

Y. Sofue, *Galactic Radio Astronomy*, Lecture Notes in Physics 935,
DOI 10.1007/978-981-10-3445-9_6

is the emissivity with L being the line-of-sight depth of the emitting region. Thus the field strength can be estimated from the synchrotron emissivity (intensity) of the source.

The brightness temperature of the sky at the galactic poles of the galaxy at $\nu = 1\,\text{GHz}$ ($\sim$ face-on brightness of the galaxy at the sun) is $T_B \sim 10\,\text{K}$. Therefore if we take a thickness of the radio disk to be 500 pc, the emissivity is

$$\epsilon \sim \nu I_\nu / L \sim 2kT_B\nu/\lambda^2 L \sim 2 \times 10^{-30}\,\text{erg s}^{-1}\,\text{cm}^{-3}. \tag{6.6}$$

From the above expressions we obtain $B \sim 1\,\mu\text{G}$ for an average value in the thick disk. If the disk is as thin as $L \sim 200\,\text{pc}$, then $B \sim 1.4\,\mu\text{G}$. From a more precise estimation the local magnetic field strength has been obtained as about $3\,\mu\text{G}$.

6.1.2 Linearly Polarized Emission

A single electron emits a polarized radio wave, whose electric vector is parallel to the direction of acceleration, thus the wave is polarized perpendicular to the lines of force of the magnetic field. For an ensemble of electrons with an energy spectrum of the form $N(E)dE \propto E^{-\beta}dE$, the total synchrotron intensity has the maximum polarization degree of

$$p_{\max} = \frac{\beta + 1}{\beta + 7/3} = \frac{-\alpha + 1}{-\alpha + 5/3}, \tag{6.7}$$

where $\beta \sim 2.4$ for the solar vicinity value and $\alpha = -(\beta - 1)/2 \sim -0.7$ is the spectral index of the emission ($I_\nu \propto \nu^\alpha$). Therefore, the maximum polarization degree from a source with an ideally aligned magnetic field is 70%. However, interstellar magnetic fields are usually not so well aligned, but are more tangled and random. Also the galactic radio emission is mixed with thermal radiation. Therefore the galactic emission is usually polarized by about $p \sim$ a few %.

In a region such as the galactic center where the radio spectrum is flat, the maximum polarization degree is smaller. For example, the radio arc has a spectral index of $\alpha \sim 0.2$, which yields the maximum polarization degree of $p_{\text{maz}} \sim 54\%$. Polarization observations at 30–50 GHz using the 45-m telescope at Nobeyama and with the Bonn 100-m telescope have shown that the arc is polarized by about 50%. This high degree of polarization indicates that the arc consists of almost perfectly aligned magnetic lines of force. Such field lines of force have indeed been observed with the very large array (VLA).

The direction of intrinsic polarization vectors is perpendicular to the magnetic field direction. Hence, if we can correct for the Faraday rotation as described below, we can obtain the direction of the magnetic field as projected in the sky.

6.1.3 Faraday Rotation

A linearly polarized radio wave can be expressed by a superposition of two circularly polarized waves whose electric vectors rotate in the clockwise and counterclockwise directions, respectively, called the right-hand (R) and left-hand (L) polarizations, or waves of ordinary and extraordinary modes. If an interstellar matter (ISM) involves both ionized gas (including a partially ionized one) and a magnetic field, the ISM is not isotropic for wave propagation, because the magnetic field defines a certain fixed direction in the medium. Because thermal electrons of the gas rotate in the same direction around the field lines of force by the Lorenz force, interaction by the R mode electric wave vector is different from that by L mode waves (Fig. 6.1).

This situation causes a Faraday effect on the propagating wave, thus the refractive indices n of the R and L waves are different:

$$1 - n^2 = \frac{1.24 \times 10^4 n_e\,(\mathrm{cm}^{-3})}{\nu^2(1 \pm \nu_H \cos\,\theta)}, \tag{6.8}$$

where θ is the angle between the line of sight and field direction, and ν_H is the Larmor frequency of the field B. Because of a phase shift of the two modes, the resultant linear polarization vector rotates as the wave propagates. This is called the Faraday rotation of linearly polarized radio waves, and the variation of the position angle ϕ of the wave is given by

$$\phi = \phi_0 + RM\lambda^2, \tag{6.9}$$

where $\phi_0 = \phi_{\mathrm{mag}} + 90°$ is the intrinsic polarization angle with ϕ_{mag} the position angle of the field lines projected in the sky. RM is called the rotation measure, and is given by

$$RM = 0.81 \int_0^x n_e B_p dx\ (\mathrm{rad\ m^{-2}}). \tag{6.10}$$

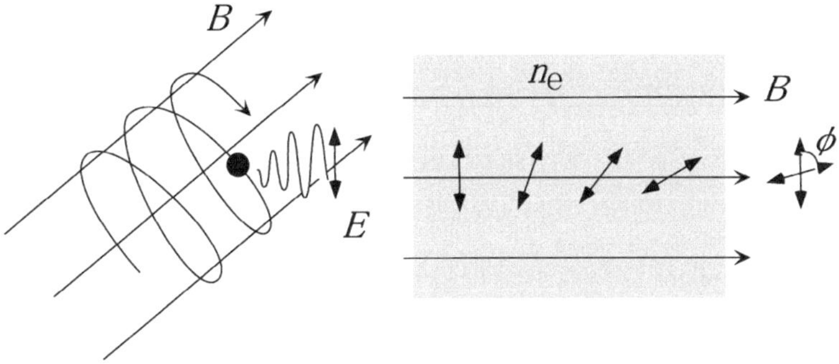

Fig. 6.1 Linear polarization and Faraday rotation

Here, n_e (in cm^{-3}) is the thermal electron density, x (in pc) is the distance along the line of sight, and $B_p = B\ \cos\theta$ (in μG) is the line-of-sight component of the field.

RM is defined as positive when the field line runs away from the observer, and *RM* is negative if the field line is toward the observer. If we observe position angles ϕ of linearly polarized radio waves at different wavelengths and plot them as a function of λ^2, we can determine $RM \sim 0.81 n_e B_p L$. If we assume a certain value for the thermal electron density n_e and the depth L, we can approximately determine the line-of-sight component of the field.

6.1.4 Determination of Magnetic Field Orientation

In galaxies and ISM, radio-emitting sources are a mixture of the synchrotron source and thermal gas. In such circumstances, the three-dimensional orientation of the magnetic field can be determined by observing synchrotron radio intensity and linearly polarized emission as follows. (a) Observed intensity I_ν of synchrotron radiation gives the field strength $B = (B_t^2 + B_p^2)^{1/2}$. (b) Intrinsic polarization angle $\phi_0 - 90°$ gives the transverse direction of the field line in the sky B_t. (c) Faraday rotation measure *RM*, which is determined from multiwavelength measurement of the polarization angle, gives the line-of-sight field component B_p.

Figure 6.2 illustrates how the Faraday rotation measure (RM) is related to the thermal electron density and magnetic field, and how the RM value is determined by multifrequency observations.

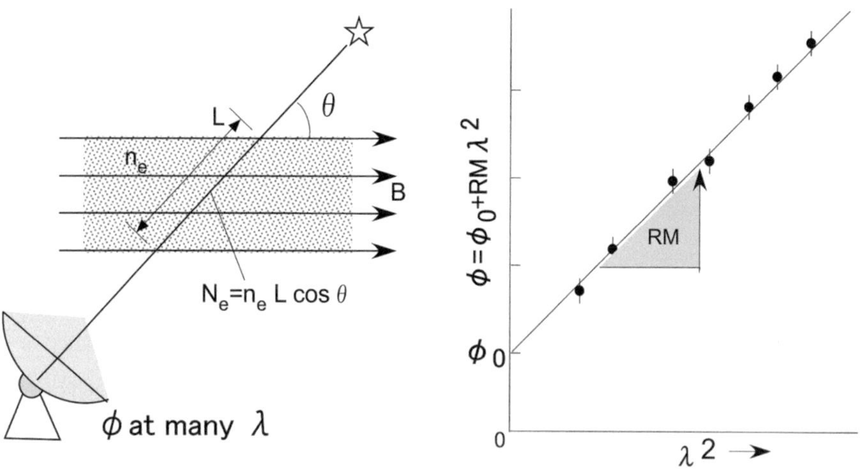

Fig. 6.2 Determination of RM by measuring polarization angle at various λ^2, which is used to estimate B by assuming the electron density n_e and line-of-sight depth L as well as the angle between the field line and line of sight

6.1.5 RM-vs-HI Density Method for Galactic Local Field

When magnetic lines of force are filled with HI gas making a string in the sky with known direction, as illustrated in Fig. 6.3, RM values plotted against HI column density can be used for estimating the field strength. This method is applied later to the local field in the Aquila Rift.

If the magnetic field is ordered, the rotation measure is simply proportional to the column density of thermal electrons N_e as $RM \propto BN_e$, and further $RM \propto N_e$ for constant B. RM may thus be related to the HI density and expected thermal electron fraction as

$$RM \simeq \Gamma x_e N_{HI} B \cos\theta, \tag{6.11}$$

where $\Gamma = 2.62 \times 10^{-19}\,\mathrm{rad\,m^{-2}\,cm^2\,\mu G^{-1}}$, N_{HI} is the HI column density in $\mathrm{cm^{-2}}$. Here, x_e is the free-electron fraction.

By representing RM by a linear function of N_{HI} as

$$RM = QN_{HI} + \mathrm{const.} \tag{6.12}$$

with

$$Q = \frac{dRM}{dN_{HI}} \tag{6.13}$$

the magnetic field strength is calculated by measuring the gradient Q as

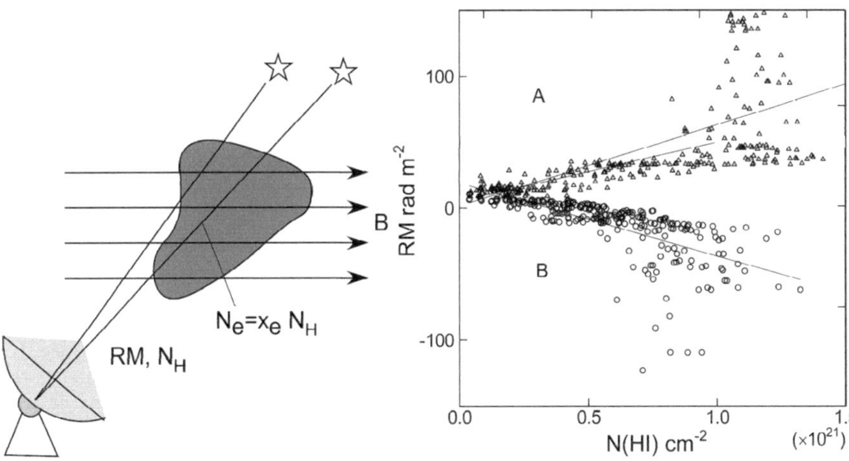

Fig. 6.3 Rotation measure plotted against HI column density in regions A and B centered on $(l, b) = (20°, 20°)$ and $(0°, 40°)$, respectively. The *straight lines* show least-squares fitting results

$$B \simeq 3.82 \times 10^{18} \frac{Q}{x_e \cos\theta} \ \mu\text{G}, \tag{6.14}$$

with Q measured in rad m^{-2} cm^2.

The free-electron fraction x_e is known to be on the order of $x_e \sim 0.08$ from measurements of dispersion measure and HI absorption toward pulsars and from UV spectroscopy of stars for the warm neutral hydrogen in the local interstellar gas. The Q value is determined by linearly fitting the plots in RM-versus-$N_{\rm HI}$. The field strength is then estimated using Q, assuming the angle between the line of sight and the field lines.

6.1.6 RM-Gradient Method

The usual RM analysis cannot be applied to regions where the magnetic field is perpendicular to the line of sight, so that $RM \sim 0$. However, for regions neighboring the perpendicular point, we can estimate the field strength by measuring the RM value gradient along the supposed field direction.

Let us consider that we observe RM as a region around the perpendicular point of a magnetic tube as illustrated in Fig. 6.4. Introducing an angle $\xi = 90° - \theta$ equation the magnetic field is related to the gradient of RM by

$$B \simeq \frac{dRM}{d\xi} \frac{1}{\Gamma x_e N_{\rm H}}, \tag{6.15}$$

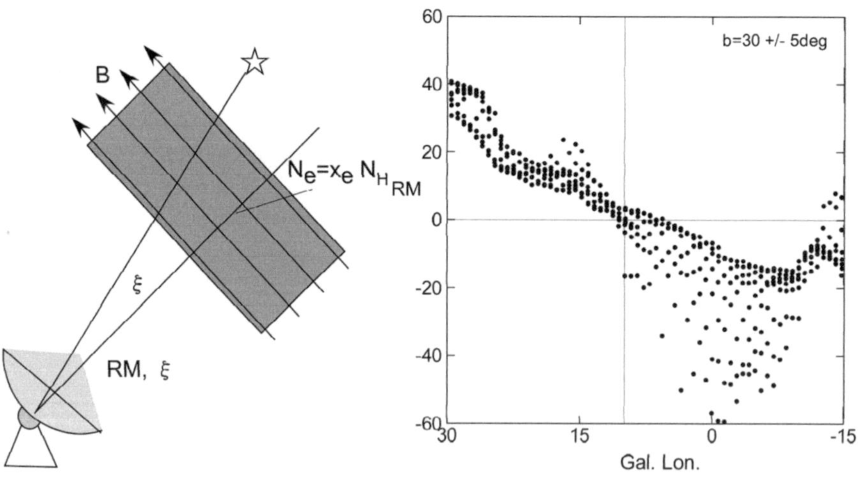

Fig. 6.4 Rotation measure plotted against angle from the perpendicular point at $(l, b) = (10°, 30°)$ along the Aquila Arch

for small ξ around the perpendicular point to the field lines. Measuring B in μG, $N_{\rm H}$ in H cm^{-2}, RM in rad m^{-2}, and ξ in degrees, this equation is rewritten as

$$B \simeq 2.2 \times 10^{20} \frac{dRM}{d\xi^{\circ}} \frac{1}{x_{\rm e} N_{\rm H}}, \tag{6.16}$$

or if $x_{\rm e} = 0.08$ as above, we obtain

$$B \simeq 2.7 \times 10^{21} \frac{dRM}{d\xi^{\circ}} \frac{1}{N_{\rm H}}. \tag{6.17}$$

This method is applied to the Aquila Rift in the next subsections.

6.2 Magnetic Fields in the Milky Way

The global magnetic structure in the Milky Way can be studied by analyzing linear polarization data of synchrotron radiation. The line-of-sight components of the galactic magnetic field can be obtained by the variation and distribution of the Faraday rotation measure of extragalactic sources in the sky, as shown in Fig. 6.5. The perpendicular magnetic directions can be known by plotting the B vectors (perpendicular to the polarization angle) in the sky, as shown in Fig. 6.6.

These polarization data of the synchrotron emission reveal projected magnetic structures integrated along the lines of sight through the entire galaxy. Magnetic orientation can also be obtained by dust polarization by far-infrared observations, as shown in Fig. 6.7, which manifest more local magnetic structures associated with dust clouds in the solar vicinity.

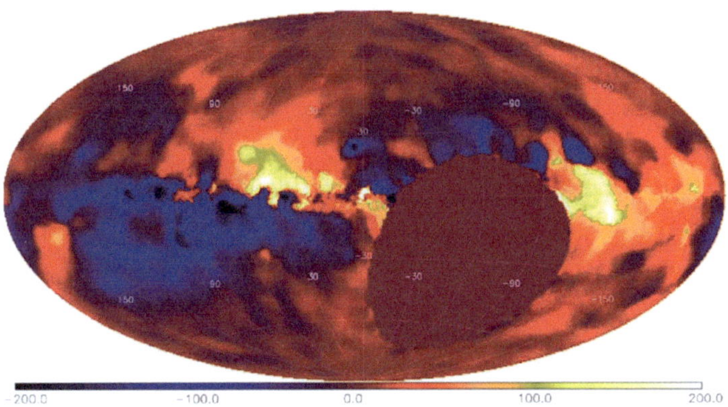

Fig. 6.5 Distribution in the sky of averaged Faraday rotation measure (RM) of extragalactic radio sources [8]

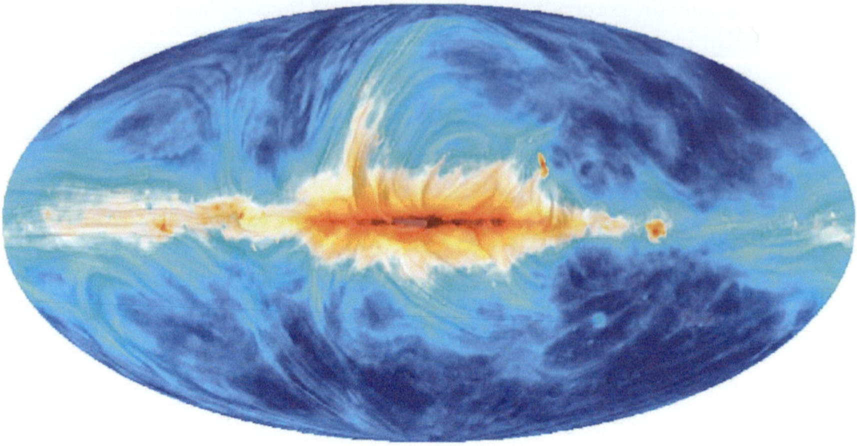

Fig. 6.6 Distribution of B-vectors by interstellar synchrotron emission at 28 GHz

Fig. 6.7 Distribution of B-vectors by interstellar dust emission [5]

6.2.1 Local Magnetic Fields: A Case for Aquila Rift

The local interstellar space in the solar vicinity at distances within a few hundred pc is filled with neutral gas clouds and hot ionized plasma. Neutral clouds are composed of neutral hydrogen and molecular hydrogen gases, emitting the HI line and molecular lines such as CO. Molecular clouds are also observed in dust emission in far-infrared wavelengths and optical extinction of starlight. The hot plasma is observed as diffuse background X-ray emission.

Because the synchrotron emission from the local region is weak compared to the galactic disk emission, magnetic field measurement by linear polarization observation is difficult. Instead, the local fields measured are by stellar polarization caused by scattering due to dust aligned perpendicular to field lines, as well as by far-infrared polarized emission by dust. Rotation measures of extragalactic sources are also used to measure magnetic fields along the line of sight through the local clouds.

The most prominent magnetized ISM phenomenon in the solar vicinity is observed in the Aquila Rift region. The Aquila Rift is a giant dark lane dividing the Milky Way by heavy starlight extinction at $l \sim 10$–$30°$, making a giant triangular shade. It extends to positive latitudes over the galactic center, and returns to the galactic plane at $l \sim 270°$. It is composed of a more extended HI arch, called the Aquila Arch, and dust filaments, drawing a giant arch in the sky (Fig. 6.8). Figure 6.9 illustrates the 3D structure of the rift, and Fig. 6.10 shows an HI map around the Aquila Arch.

The Arch is associated with magnetic fields aligned along the ridge as inferred from starlight polarization and FIR polarized dust emission (Fig. 6.9). The root region at $l \sim 20$–$30°$ apparently coincides with the root of the North Polar Spur (NPS), but the NPS is a galactic center object as discussed in the previous chapter, and is a separate structure from Aquila.

Polarization of dust emission and starlight polarization show that the magnetic fields in the Aquila Arch run parallel to its ridge. The line-of-sight direction of the magnetic field can be obtained from the distribution of the Faraday rotation measure.

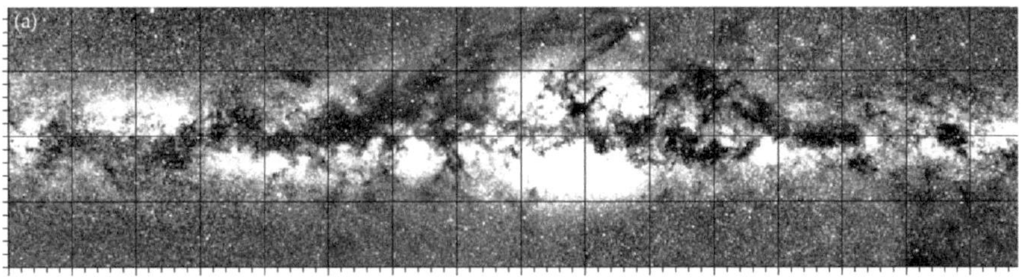

Fig. 6.8 The Aquila Arch in dark clouds associated with molecular clouds [2] in the $|l| \leq 90°$ region around the GC

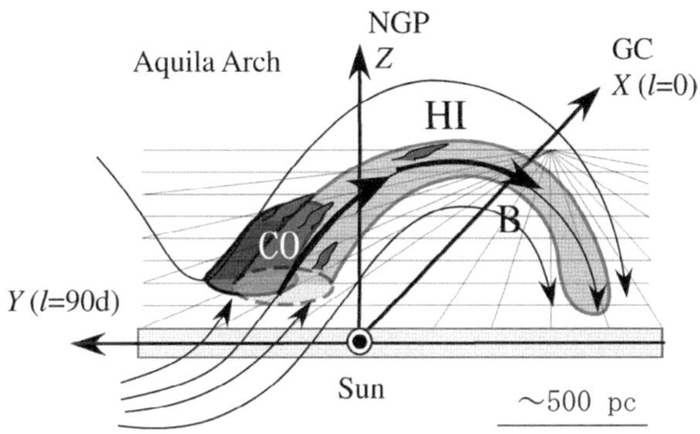

Fig. 6.9 Coordinates used in the analysis, and illustration of 3D structure of the Aquila Arch

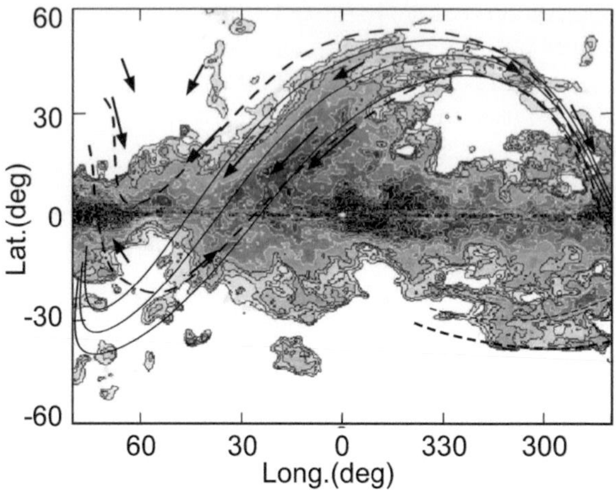

Fig. 6.10 Magnetic lines of force drawn after 2D simulation of Parker instability superposed on the magnetized HI arch in the Aquila Rift

Figure 6.10 shows flow lines of the magnetic fields and their line-of-sight direction, indicating the 3D magnetic lines of force.

6.2.2 *Energetics of the Aquila Rift*

Using HI intensity distribution and the linear extent, the mass of the HI arch in the Aquila Rift is estimated to be $M_{\rm HI} \sim\sim 1.8 \times 10^5 M_\odot$. From the CO-line intensities and linear extent, the molecular mass of the Aquila-Serpens molecular complex is estimated to be $M \sim 3 \times 10^5 M_\odot$. The gravitational energy to lift the HI gas to height z from the galactic plane is estimated by

$$E_{\rm g} = \int M k_z dz \simeq M k_z z, \tag{6.18}$$

where k_z is the vertical acceleration. At the mean height of the HI arch of $z \sim Z_0/2 \sim 250\,\rm pc$, the acceleration is $k_z \simeq 5 \times 10^{-9}\,\rm cm\,s^{-2}$. Thus the gravitational energy of the HI Arch is estimated to be $E_{\rm g,HI} \sim 1.4 \times 10^{51}$ erg.

The magnetic energy in the HI Arch is estimated by assuming a magnetic tube of length $L_1 \sim L/2$ and width L_2 with constant field strength as $E_{\rm mag} \simeq B^2/8\pi L_1 L_2^2$, Inserting the measured strength $B \sim 10\,\mu\rm G$ and the size $L_1 \sim 1\,\rm kpc$ and $L_2 \sim 100\,\rm pc$, we obtain a magnetic energy of $E_{\rm mag} \sim 1.2 \times 10^{51}$ erg, which is comparable to the gravitational energy. Thus we learn that the magnetic energy is sufficient to lift the arch to its present height by magnetic pressure.

6.2.3 *Parker Instability in the Aquila*

As to the inflation mechanism of the Aquila Arch we may consider Parker instability. The magnetic energy of the Arch is known to be comparable to the total gravitational and kinetic energy of the HI and H_2 gases. If we assume the pressure equilibrium between cosmic rays and the magnetic field, the pressure is strong enough to raise the gas against the gravitational force, therefore the condition for the growth of Parker instability is satisfied. Simulations showed that the instability can grow either in symmetric or asymmetric waves with respect to the galactic plane. In Fig. 6.10 we reproduce the magnetic field lines calculated by 2D MHD simulation superposed on intensity maps of local HI and CO gases.

A possible scenario of formation of the Aquila Rift is as follows.

(1) The galactic disk about $\sim$250 pc in the GC direction was penetrated by an ordered magnetic field parallel to the solar circle, and was compressed by a galactic shock wave to increase the density and magnetic field.
(2) A Parker instability took place along the shocked compressed high-density arm with a strong magnetic field, and the gas was inflated by an expanding magnetic tube to a height $\sim$500 pc with a wavelength of $\sim$2.5 kpc.
(3) It occurred at a timescale of $t \sim \text{height}/v_A \sim 5 \times 10^7$ y, where $v_A \sim 10\,\text{km s}^{-1}$ is the Alfvén velocity for $B \sim 10\,\mu\text{G}$ and $n_H \sim 5\,\text{H cm}^{-3}$.
(4) The inflated HI gas slipped down along the field lines toward the Aquila region, where the gas was compressed by dynamic pressure of the falling motion to form the HI Arch along the field lines.
(5) The most strongly compressed gas in the accumulating region at the root of the field lines, a phase transition from HI to H_2, occurred to form the Aquila-Serpens molecular complex.
(6) The molecular complex and HI Arch are separated by a thin layer of phase transition, where the two gas phases avoid each other.
(7) The entire Parker instability took place in a sinusoidal wave, with the counter arch observed at negative latitudes around $(l, b) \sim (300°, -30°)$.

6.3 Magnetic Fields in Disk Galaxies

6.3.1 *RM in Disk Galaxies*

The magnetic field configuration in disk galaxies can be determined by the Faraday rotation observations, which give us information about the transverse and line-of-sight components of the field as described above. A widely adopted method to determine magnetic field configuration is called the *RM* versus Θ (azimuthal angle) method.

In this method we plot the value of *RM* (or equivalently $\Delta\phi = \phi - C$ where C is an arbitrary constant) as a function of Θ, where Θ is the azimuthal angle along a circle on the disk plane of the galaxy at a fixed radius (Fig. 6.11).

If the magnetic field of the galaxy disk is ringlike or axisymmetric (ASS; as if the field apparently has nonzero divergence), *RM* shows a single sinusoidal variation with Θ. If the field has a bisymmetric spiral (BSS) configuration, in which the field line flows in from one edge of the galaxy and flows out from the opposite edge, the *RM* variation is double-peaked sinusoidal along an annulus ring.

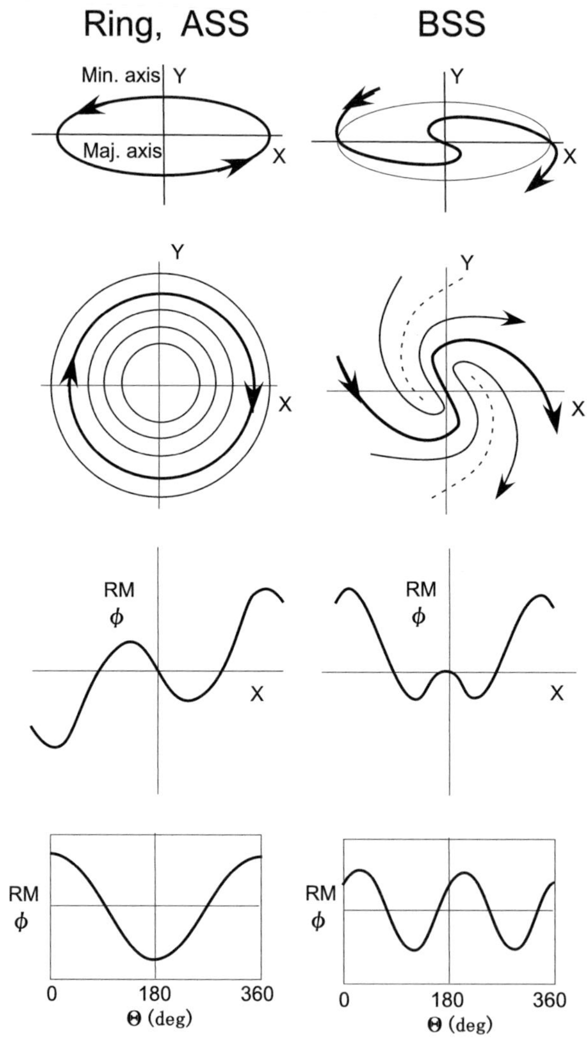

Fig. 6.11 Variation of *RM* with Θ according to the field configurations: ring or axisymmetric (ASS), and bisymmetric spiral (BSS) [6]

A simpler analysis can be obtained if the RM or ϕ variation is measured along the major axis. For a ring or ASS field, the variation is anti-symmetric with respect to the galactic center, as shown in Fig. 6.11. On the other hand, the variation is symmetric for a BSS field.

There have been a number of galaxies for which the field configurations have been determined in this way. The galaxy, M51, M81, NGC 4258, and so on have a BSS field; M31 has a ring field, and NGC 6946 appears to show a spiral field with an apparent nonzero divergence.

6.3.2 *Ring Field in M31*

The most typical case of a ring field is observed in the Andromeda galaxy (M31). Figure 6.12 shows offsets of position angle of linear polarization ϕ at 2.7 GHz from the annulus ring as projected in the sky plotted against the azimuthal angle.

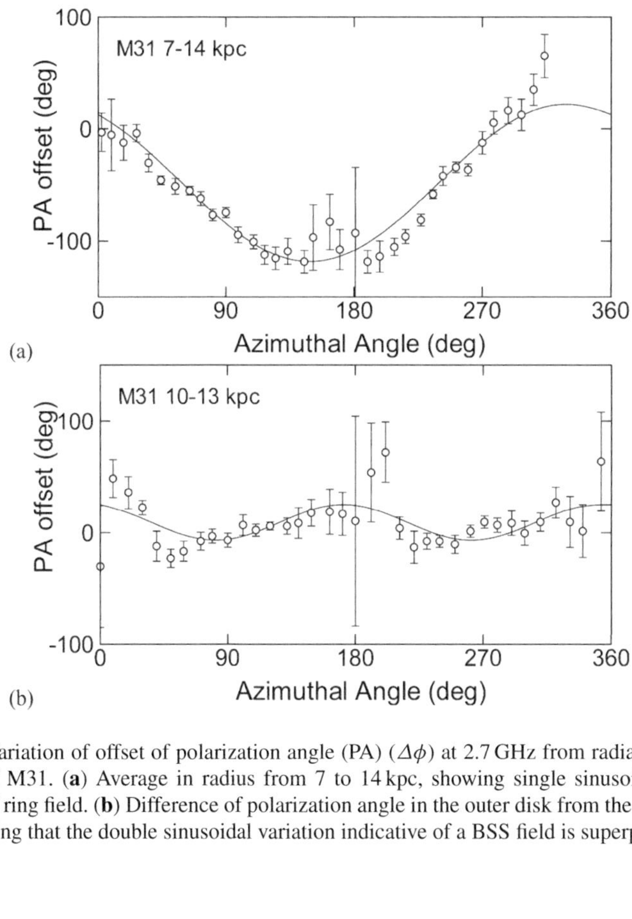

Fig. 6.12 Variation of offset of polarization angle (PA) ($\Delta\phi$) at 2.7 GHz from radial direction in the plane of M31. (**a**) Average in radius from 7 to 14 kpc, showing single sinusoidal variation indicative of ring field. (**b**) Difference of polarization angle in the outer disk from the sinusoidal fit in (**a**), showing that the double sinusoidal variation indicative of a BSS field is superposed

If there is no Faraday rotation in the galactic disk, the polarization angle points the direction perpendicular to the magnetic field, and hence the offset from the field direction must remain constant. However, the figure shows a single sinusoidal variation with Θ. This indicates that the polarization vectors are rotated from the intrinsic directions, and their amplitudes are proportional to RM in the disk. The sinusoidal variation of $\Delta\phi$, and therefore of RM, indicates that the field orientation is a ring.

In the case of M31, the intrinsic polarization angle after correcting for the Faraday rotation is shown to be perpendicular to the azimuthal ring, indicating that the magnetic field is parallel to the ring. The ring field is a particular case of the ASS field when the pitch angle of the field lines is zero.

6.3.3 *BSS Field in M51*

A typical case of a bisymmetric spiral magnetic field is observed in the spiral galaxy M51, as shown in Fig. 6.13. The figure shows observed B-vectors of linear polarization at 5 GHz, pointing the directions of the magnetic field projected in the sky that show a spiral configuration.

In addition to the B-vector distribution, a map of RM is available in this galaxy from multiwavelength measurements of linear polarization. The RM value sign indicates the line-of-sight field direction. Combining the RM sign and the inclination of the plane of the galactic disk, we can determine the 3D direction of the magnetic field. The thus- determined field directions of M51 are shown by red and blue arrows in Fig. 6.13, with red arrows indicating $RM > 0$ (field toward the observer) and blue $RM < 0$ (away from the observer).

Using the B-vector distribution and 3D field directions by red and blue arrows, we may draw a possible field configuration as shown in the bottom panel of Fig. 6.13.

6.3.4 *Vertical Fields in Spiral Galaxies*

In addition to the disk fields, numerous evidence for vertical magnetic fields has been observed, which may be more or less related to the activities of the disk and the nucleus.

6.3.4.1 Vertical Structures from the Galactic Disks

The radio-continuum maps of the Milky Way show vertical structures emerging from the galactic plane. Their appearance parallel to each other suggests their coherent origin, likely driven by magnetic lines of force emerging normal to the disk plane.

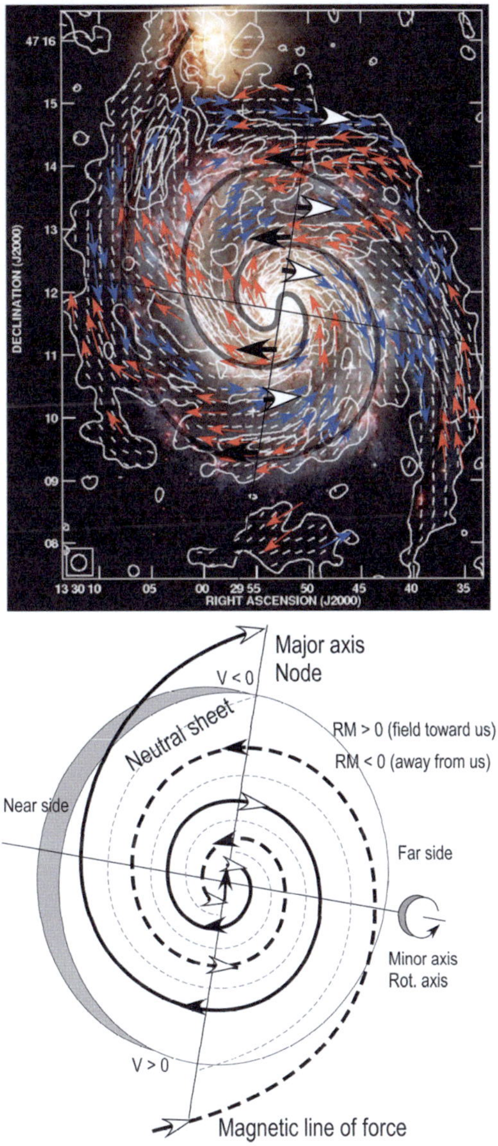

Fig. 6.13 (*Top*) B-vectors [3] on an image of M51 (HST) and major spiral arms in *grey line*. *Red* and *blue arrows* are line-of-sight field directions determined by RM signs with *red* showing the field toward the observer and *blue* away from the observer. *Big black* and *white arrows* show representative directions of the field on the major axis (node of the disk) inferred from the radial velocity field. (*Bottom*) Illustration of a possible BSS magnetic line of force in M51

External edge-on galaxies often show vertical dust lanes. Long, coherent, thin filaments running normal to the disk of some dust-rich spirals suggest the existence of a large-scale vertical field running across the disk plane. These vertical structures are found at radii of a few kpc and run for more than a kilo parsec in the halo.

6.3.4.2 Vertical Magnetic Fields in the Galactic Nuclei

In the central region of the galaxy direct evidence for a vertical field has been found with high-resolution and/or polarization observations of synchrotron radio emissions. A large number of straight filaments extending for a hundred pc scale run almost perpendicular to the disk plane near the arc and are well understood as the trace of a magnetic field running vertical to the disk. Some parts of these structures show strong polarization and Faraday rotation directly showing the poloidal magnetic field.

Polarization observations of the nuclear radio source in M31 shows the magnetic field orientation perpendicular to the major axis. Because the galaxy is nearly edge-on, this may be attributed to a poloidal magnetic field in the nucleus. In addition to M31, however, no obvious magnetic structures are known for nuclei in external galaxies.

6.3.4.3 Vertical Ejections from Nuclei of Galaxies

Nuclei of spiral galaxies often reveal jetlike features emerging perpendicular to the disk plane and/or central radio sources elongated perpendicular to the major axis. These ejection features may be the manifestation of a vertical field running across the nucleus.

6.3.5 Magnetic Fields in the Galactic Halo

Observational data for halo fields are still crude. A few edge-on galaxies such as NGC 4631 show an extended nonthermal radio halo. The field strength from the radio emissivity is estimated to be a few μG. There are many galaxies for which no evidence of a radio halo is seen.

Magnetic fields in the halo of the galaxy can be derived from *RM* analysis for external radio sources and pulsars. If we plot *RM* against cot b for radio galaxies and quasars, the upper envelope of the plot can be fitted by a relation,

$$|RM|_{\rm RG,Quasar} \simeq 30 \cot |b| \ ({\rm rad\,m^{-2}}). \tag{6.19}$$

On the other hand if we plot the same for pulsars, we obtain

$$|RM|_{\rm Pulsar} \simeq 10 \cot |b| \ ({\rm rad\,m^{-2}}). \tag{6.20}$$

The difference between the coefficients for the two plots is considered to be caused by Faraday rotation in the space above a disk of thickness of about 500 pc in which most pulsars are distributed; namely it may be due to a halo beyond a few hundred pc from the galactic plane:

$$|RM|_{\mathrm{Halo}} \sim 20 \cot |b|. \tag{6.21}$$

If we take an electron density and a thickness of the halo to be approximately $10^{-3}\,\mathrm{cm}^{-3}$ and a few kpc, respectively, the field strength in the halo is estimated to be a few μG, about the same as in the disk.

6.4 Origin of Magnetic Fields

To create a magnetic field in galaxies, two major mechanisms are considered.

(1) The dynamo mechanism by interstellar turbulence and differential rotation of a gas disk amplifies initially weak fields. A ring field with tight reversal from one ring to another is created.
(2) A primordial field trapped in a primeval galaxy is wound up by the differential rotation of the disk, and the spiral configuration is maintained by the dynamo mechanism between the disk and halo. The BSS configuration in spiral galaxies can be more naturally understood on the basis of the primordial origin hypothesis. The primordial hypothesis can also explain the ring configuration as the result of a reconnection of field lines wound up in the inner region of the galaxy.

6.4.1 Dynamo Mechanism for Ring and Turbulent Fields

A magnetic field is practically frozen into the interstellar gas because of the large magnetic Reynolds number and small resistivity of the fluid caused by the extremely large-scale size of the region considered compared to the atomic resistivity scales. In this case a magnetic field line obeys an induction-and-dynamo equation, expressing the time evolution of B by

$$\frac{\partial \boldsymbol{B}}{\partial t} = \mathrm{rot}(\boldsymbol{u} \times \boldsymbol{B}) + \eta \nabla^2 \boldsymbol{B} + \mathrm{rot}(\Gamma \boldsymbol{B}), \tag{6.22}$$

and

$$\mathrm{div}\boldsymbol{B} = 0, \tag{6.23}$$

where $\boldsymbol{u}$ is the velocity of gas in galactic differential rotation, η is the magnetic dissipation coefficient (diffusion constant) caused by the turbulent motion, and Γ is the coefficient of dynamo action having the dimension of velocity.

The first term represents Maxwell's induction relation between the magnetic field and electric current $\boldsymbol{j}$, when the gas flow velocity is assumed to be proportional to the current as $e\boldsymbol{u} \propto \boldsymbol{j}$. This relation holds for a system with sufficiently large conductivity, and is the case in astrophysical bodies, where ionization degree, even in neutral ISM, is high enough that the gas is regarded to be an infinitely conductive body. The second term including η represents dissipation of the magnetic energy by an induced current in a resistive medium. Although the atomic resistivity is negligible, anomalous resistivity becomes efficient by interstellar turbulence, which depends on the Alfvé velocity.

The third term including Γ represents the dynamo action induced by various kinds of amplification processes such as the $\alpha - \omega$ dynamo combined with Parker instability and magnetorotational instability (MRI).

Here we consider interstellar and galactic magnetic fields. Those in and around stars, pulsars, and active nuclei are beyond the scope of this chapter. In Fig. 6.14 we illustrate some typical ideas to create and amplify galactic magnetic fields by the dynamo mechanism. By Parker instability, a part of the galactic field inflates toward the halo. By magnetorotational instability, perturbation of field lines occurs in the radial direction.

6.4.1.1 Parker Dynamo

In the Parker dynamo, perturbed magnetic field lines inflate in the halo direction forming an Ω shape, and expand due to cosmic-ray pressure [4]. By the buoyancy against the z-directional gravity by the galactic disk, the inflation and expansion

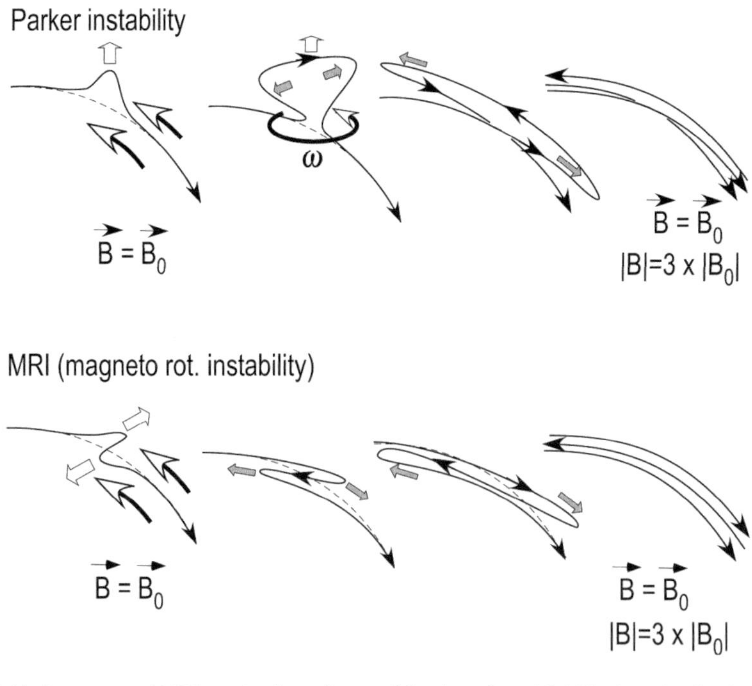

Fig. 6.14 Dynamo and MRI mechanisms for amplification of seed field in the galactic plane

are accelerated resulting in a large bubble of the magnetic field at high altitudes. Because the angular momentum of the bubble is conserved, it begins to rotate in the opposite direction from the galactic rotation, where the differential rotation of the disk and local epicyclic motion are the driving force.

The expanded edges of the bubble then suffer from the galactic differential rotation, and the field lines are stretched along the original field. These stretched magnetic field lines of force are now three times the original field line, and hence the field strength increases by a factor of three. However, the average field vector does not increase at all. Thus the mechanism can act to increase local field strengths, but cannot increase the average field intensity and cannot change the global field configuration.

6.4.1.2 MRI Dynamo

The magnetorotational instability is a similar mechanism to Parker instability, but the perturbed field lines grow in the radial as well as in the z-directions of the disk, and the shear motion by the differential disk rotation is the driving force [1]. The amplified wavy field lines suffer from the differential rotation of the disk, and are stretched in the direction of the rotation at the outer radius, and in the opposite direction at the inner radius. As in the case of the Parker dynamo, local field strength increases by a factor of three per one perturbed field line, but the global field configuration does not change.

6.4.1.3 Turbulent Dynamo

Even without the Γ term in Eq. (6.22), magnetic fields are more locally amplified through the term rot$(\boldsymbol{u} \times \boldsymbol{B})$ because of the vortices in the ISM as is usually the case in dynamic conditions such as cloud motion, rotation, collisions, and/or turbulence. Figure 6.15 illustrates the mechanism.

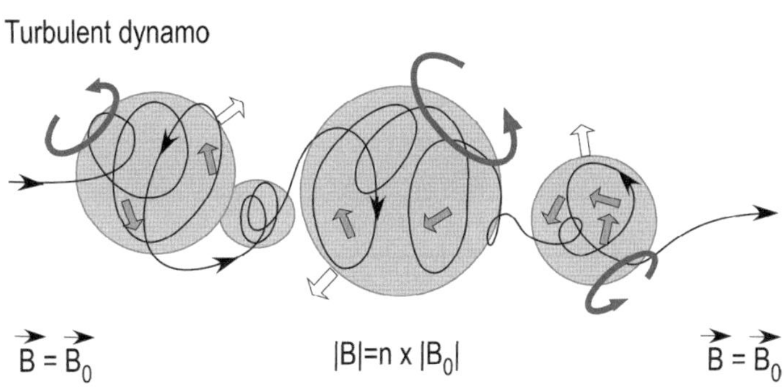

Fig. 6.15 Dynamo amplification of seed field by interstellar clouds in rotation and turbulent motions

This mechanism is efficient in local interstellar space, where strong fields of scales on the order of cloud sizes can be created. However, it does not affect the global field configuration.

6.4.2 Primordial Formation of BSS, Ring/ASS, and Vertical Fields

The large-scale BSS configuration of a galactic magnetic field cannot be created by the dynamo mechanism. It is considered to have been created during the formation of primeval galaxies. On this primordial-origin hypothesis, the present BSS field is interpreted as the fossil of an intergalactic field wound up by the primordial galaxy disk during its formation and contraction (Figs. 6.16, 6.17, and 6.18).

6.4.2.1 BSS Field

The field is amplified by winding up and reaches strength with the magnetic energy density becoming comparable to the interstellar turbulent kinetic energy.

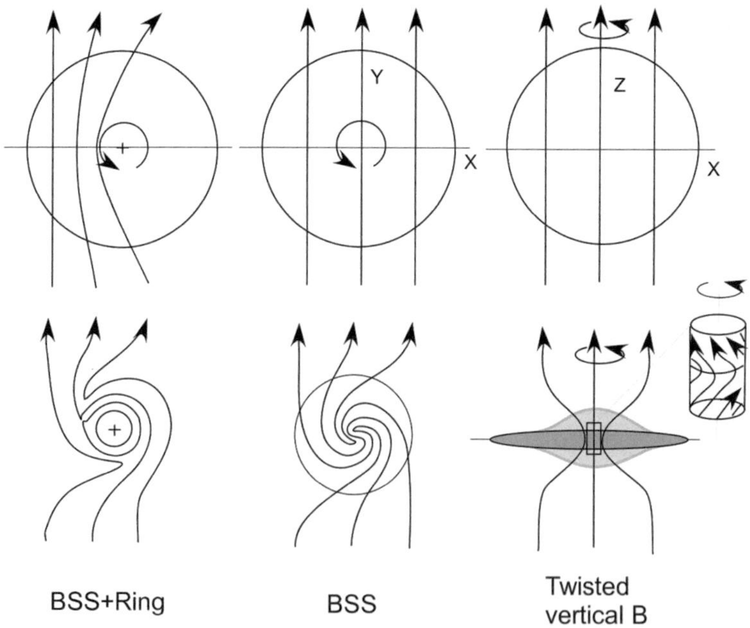

Fig. 6.16 Primordial origin of BSS and ring fields in galaxies and vertical twisted field in the galactic center

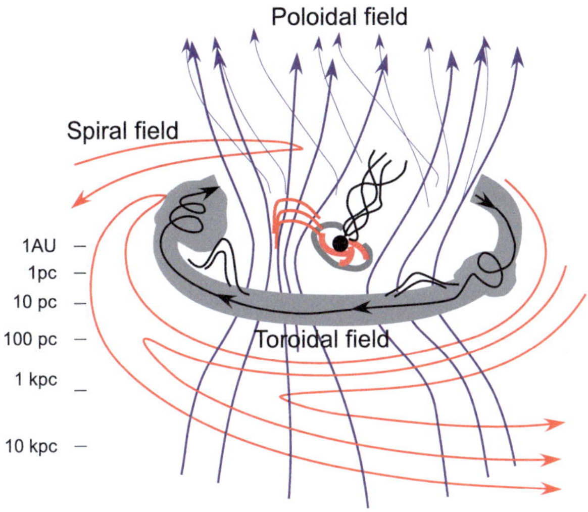

Fig. 6.17 Schematic magnetic view of the galaxy. The scale is pseudo-logarithmic in order to draw inner to outer features simultaneously

Then the field is maintained in a steady BSS configuration. Overamplified fields by differential rotation are thrown away from the disk by an inverse induction-dynamo mechanism.

6.4.2.2 Ring Field

Even a ring field could be produced from the primordial field, if there were asymmetry in the protogalactic disk with respect to the galaxy center or there were motion of the galaxy through intergalactic space.

6.4.2.3 Vertical Field

The perpendicular component of the primordial field to the galactic plane is also trapped by the rotating disk. However, the vertical field component can neither escape from the disk nor be dissipated by the dynamo mechanism. The vertical fields, therefore, accumulate in the central region according to the contraction of the galaxy disk during galaxy formation. Thus a strong vertical magnetic field is created in the galactic center.

6.4.2.4 ASS Field

An ASS field is created during the winding of the vertical component, as illustrated in Fig. 6.19. The twisted vertical field is bent toward the galactic disk, and a

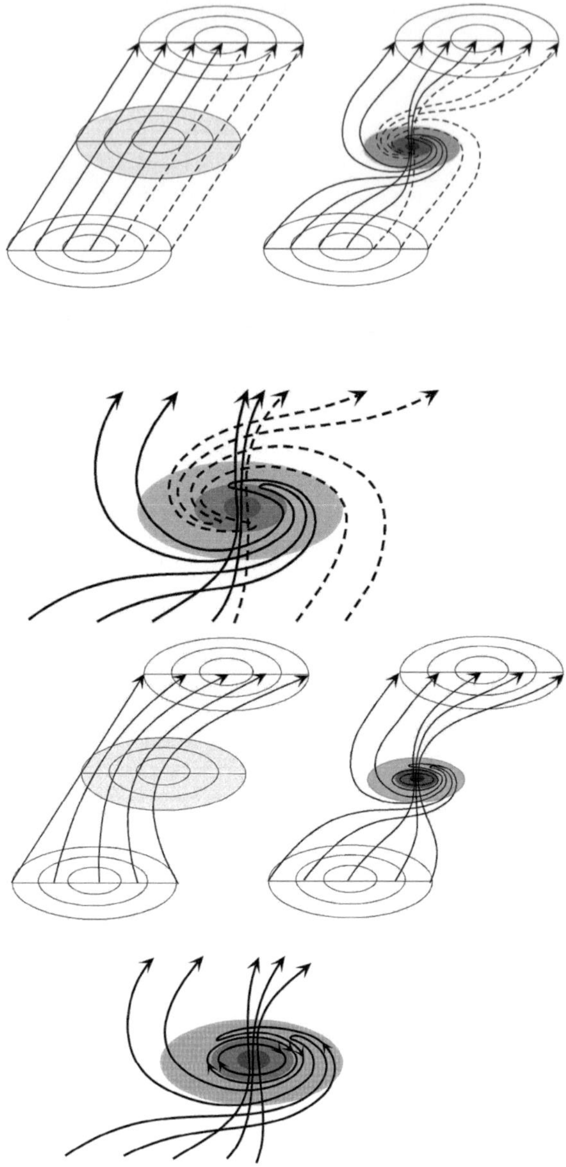

Fig. 6.18 Primordial origin of BSS and vertical magnetic fields. The ring field was created when the protogalaxy was moving in a uniform cosmic field

spiral configuration is realized. The spirally twisted field directions point coherently outward at one side of the disk, and inward at the other side, making the ASS configuration. Note, however, that the field direction parallel to the disk plane reverses from one side to the other. Figure 6.20 shows a result of numerical computation of the formation of the ASS field by magnetohydrodynamical (MHD) simulation.

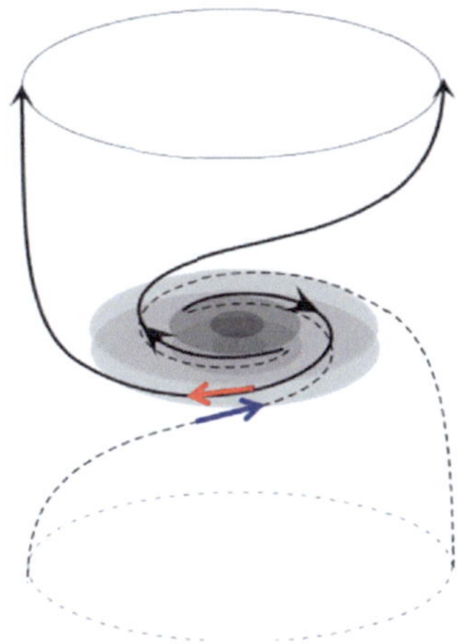

Fig. 6.19 ASS field with reversal about the plane is created by twisted vertical field

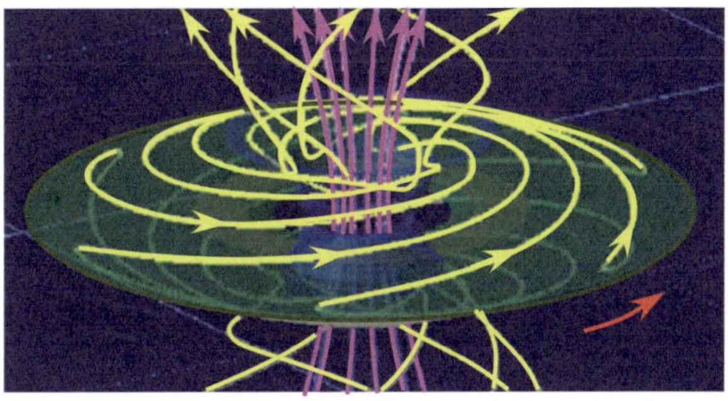

Fig. 6.20 MHD simulation of the origin of bisymmetric spiral magnetic field in the disk and vertical field in the galactic center [7]

6.4.3 *Vertical/Poloidal Magnetic Field*

It is natural that a large-scale field component parallel to the rotation axis existed, when a galaxy formed. This field component is also trapped by the primeval gas sphere [phase I]. Because the disk radius is large enough and the diffusion time is longer than the galaxy evolution time, the vertical field is almost frozen into the disk gas. The vertical field then follows an evolution as described below.

Starting from a uniform gas sphere (disk) and initial star formation in a proto-galaxy, an exponential disk is realized by the viscosity-driven angular momentum transfer and ongoing star formation. Because the field is frozen into the gas, the magnetic flux conservation results in a radial distribution of the field strength obeying the exponential law, provided the initial field is uniform. In the central region the gas density attains an excess by an order of magnitude over the value given by a simple exponential disk, and the field strength is correspondingly high. The initial star formation then finishes when the gas is fed into stars and the density decreases to a certain threshold value, after which the magnetic field is no longer frozen into the stellar disk.

6.4.3.1 Vertical Field in Galactic Nuclei

The vertical magnetic field is then frozen into the gas left behind by the initial star formation. At this stage the gas may have a constant threshold density below which the initial star formation did not take place, and shares a few percent of the total mass. The "interstellar" gas then follows its own evolution governed by the density wave shock, cloud-cloud collisions, and star formation. Through the shock- and viscosity-driven inflow the gas accretes toward the center. In the central region a bar-induced shock will enhance the accretion. Because the diffusion time of the vertical field is shorter than the dynamic timescale, this results in a formation of a strong vertical field in the center. In the early universe when galaxy formation took place, galaxy-galaxy collision will have been frequent and the tidal encounter may have enhanced the bar-induced inflow of gas, and therefore a strong vertical field near the center.

If the present intergalactic or intracluster magnetic field is on the order of $10^{-9 \sim -10}$ G, the vertical field strength in the central 1 kpc of normal spiral galaxies is expected to be on the order of mG, which dominates in the central region. On the other hand, spiral fields within the disk are weaker in the central few kpc, but dominate in the outer disk.

6.4.3.2 Loss of Angular Momentum by Vertical Field

The interstellar gas in turn suffers from magnetic torque from the vertical field which is twisted by the galactic rotation with both its sides anchored to intergalactic space at large scale. The timescale with which the rotating gas element loses angular momentum is given by $\tau = Vr/(B^2/4\pi\rho)$, where V, r, B, and ρ are the rotation velocity, radius, magnetic field strength, and gas density, respectively. The time is then calculated to be $\tau \sim 10^{11}$, 10^8, and 10^5 yr, respectively, at $r = 5$, 1, and 0.1 kpc. This shows that the accumulation of the vertical field is accelerated by the magnetic-torque/angular-momentum-loss mechanism.

6.4.4 *Cosmic Jets*

According to the scenario about accumulation and amplification of vertical field component in protogalaxies, nuclei of these galaxies had a strong vertical field. This particularly applies in stages when the condensation and density enhancement of the primeval galactic disk was prominent.

This mechanism explains many of the observed vertical radio features with active galactic nuclei, radio galaxies, and quasars. The vertical magnetic field in the nucleus is twisted, and the twist accelerates a screwing outflow of gas, forming a helical magnetic jet. Long energetic jets from Active Galactic Nucleus and quasars

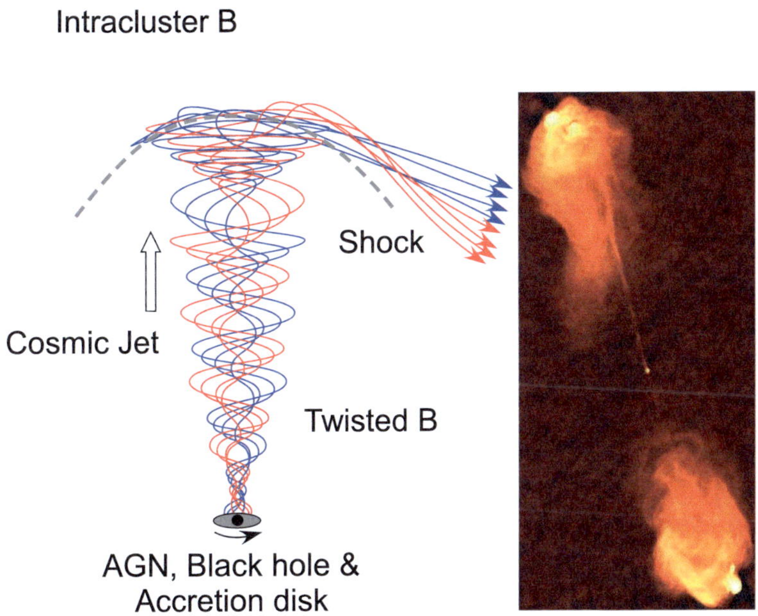

Fig. 6.21 [*Left*] Cosmic jet driven by twisting magnetic field by accretion disk around central supermassive black hole of an AGN. [*Right*] Cyg A, a strong radio source ejecting a bipolar jet extending for ± 70 kpc (figure size). The jet encounters intergalactic matter and produces bow shocks making gigantic radio lobes (Image courtesy of NRAO/AUI, http://images.nrao.edu/110)

are thought to be the manifestation of such an intensive accretion of vertical magnetic fields around the nuclear supermassive black holes of galaxies (Fig. 6.21).

6.5 Magnetic Fields in Cosmological Scale

6.5.1 Intracluster Magnetic Fields

Clusters of galaxies are associations of 10^3–10^4 galaxies embedded in high-temperature plasma of $T \sim 10^7$ K and total cluster mass of $\sim 10^{12-13} M_\odot$, including the dark matter and plasma. Clusters are a diffusely extended, but gravitationally bound system, with their extent $L \sim 1$ to several Mpc. Their sizes are represented by core radii, within which half of the luminosity or the mass is included, and is on the order of several hundred kpc to $L_{\text{cluster core}} \sim 1$ Mpc.

Magnetic fields in clusters of galaxies are measured by the same methods employed for galactic magnetic fields. The total strength of a magnetic field is estimated from the synchrotron (nonthermal) radio emissivity using Eq. (6.4), assuming the equipartition between magnetic and cosmic-ray energy densities. The emissivity is estimated using radio continuum brightness and the line-of-sight depth the same as the cluster size.

The cluster's magnetic field is also measured by observing Faraday rotation measures of radio sources behind and/or inside the cluster. Thereby, thermal electron density is estimated using the emission measure for thermal X-ray intensity of the hot plasma. If the magnetic field is ordered, RM is related to the field strength B by Eq. (6.10). If the field directions are tangled, or random, the rotation measure is related to the field strength as

$$RM = 0.81 n_e B_{\text{cluster:random}} \sqrt{L \Delta L}. \tag{6.24}$$

Here, L is the line-of-sight extent (size) of the cluster, and ΔL is the scale size of turbulent eddies in the random field. The magnetic strength B is measured in μG, RM in rad m^{-2}, L in pc, and N_e in cm^{-3}. The turbulent eddy size is not directly measurable, and is often replaced by the scale length of incoherent variation sizes of the field direction if it is measured by polarization angles.

From radio continuum and linear polarization observations of a cluster's nonthermal radio emission and Faraday rotation measurements of background and cluster radio sources, the field strength has been measured to be on the order of $B_{\text{cluster}} \sim 1\,\mu$G, and the core fields are as strong as $B_{\text{cluster core}} \sim 3$–$5\,\mu$G.

6.5.2 *Cosmological Magnetism*

The precise values and upper limits of intergalactic and cosmological-scale magnetic fields are still controversial. Such large-scale magnetic fields can in principle be measured by the same methods as used for galaxies and clusters of galaxies. Faraday rotation measures of radio sources with known red-shift z are the major tool for such purposes by analyzing RM values as a function of z.

Existence, or nonexistence of the largest scale magnetic field affecting the geometry of the universe would be a crucial subject for cosmology. If there is a uniform magnetic field running across the universe, it would be crucial for discussing the principles of cosmology, particularly regarding the isotropy principle.

Any celestial objects in the universe detected and observed thus far have exhibited anisotropy such as the rotation and the major or minor axes, which are often represented as the north and south of the object. North and south are often the directions of the penetrating polar magnetic fields as observed in planets, stars, galactic disks, and circumnuclear disks. Namely, all celestial objects have N and S, and are associated with penetrating magnetic fields. The question of whether the universe is an exception will be an intriguing subject for cosmology in the future.

References

1. Balbus, S.A., Hawley, J.F.: A powerful local shear instability in weakly magnetized disks. I – linear analysis. II – nonlinear evolution. ApJ **376**, 214 (1991) [MRI]
2. Dame, T.M., Hartman, D., Thaddeus, P.: The Milky Way in molecular clouds: a new complete CO survey. ApJ **547**, 792 (2001) [CO]
3. Fletcher, A., Beck, R., Shukurov, A., Berkhuijsen, E.M., Horellou, C.: Magnetic fields and spiral arms in the galaxy M51. MNRAS **412**, 2396 (2011) [M51 magnetic field]
4. Parker, E.N.: Cosmical Magnetic Fields. Oxford University Press, New York (1979)
5. Planck Collaboration, et al.: Planck intermediate results. XXI. Comparison of polarized thermal emission from Galactic dust at 353 GHz with interstellar polarization in the visible. AA **576**, A104 (2015)[Dust polarization]
6. Sofue, Y., Fujimoto, M., Wielebinski, R.: Global structure of magnetic fields in spiral galaxies. ARAA **24**, 459 (1986) [Magnetic field in galaxies]
7. Sofue, Y., Machida, M., Kudoh, T.: The primordial origin model of magnetic fields in spiral galaxies. PASJ 62, 1191 (2010) [MHD of BSS field]
8. Taylor A.R., Stil J.M., Sunstrum C.: A rotation measure image of the sky. ApJ **702**, 1230 (2009) [RM of radio sources]

Erratum to: Galactic Radio Astronomy

Yoshiaki Sofue

Erratum to:
Y. Sofue, *Galactic Radio Astronomy*, Lecture Notes in Physics 935, https://doi.org/10.1007/978-981-10-3445-9

The original version of this book was published without printing chapter abstract. The abstracts for each chapter are updated in the front matter in the current version of the book.

The updated online version of the book can be found at
https://doi.org/10.1007/978-981-10-3445-9

Y. Sofue, *Galactic Radio Astronomy*, Lecture Notes in Physics 935,
DOI 10.1007/978-981-10-3445-9_7

MIX
Papier aus verantwortungsvollen Quellen
Paper from responsible sources
FSC® C105338

If you have any concerns about our products,
you can contact us on
ProductSafety@springernature.com

In case Publisher is established outside the EU,
the EU authorized representative is:
Springer Nature Customer Service Center GmbH
Europaplatz 3, 69115 Heidelberg, Germany

Printed by Libri Plureos GmbH
in Hamburg, Germany